How to Solve Word Problems in Geometry

How to Solve Word Problems in Geometry

Dawn B. Sova, Ph.D.

McGraw-Hill
New York San Francisco Washington, D.C. Auckland Bogotá
Caracas Lisbon London Madrid Mexico City Milan
Montreal New Delhi San Juan Singapore
Sydney Tokyo Toronto

Library of Congress Cataloging-in-Publication Data

Sova, Dawn B.
How to solve word problems in geometry / Dawn B. Sova.
p. cm.
Includes index.
ISBN 0-07-134652-X
1. Geometry—Study and teaching. 2. Word problems (Mathematics)
I. Title.
QA461.S68 1999
516—dc21 99-37603
CIP

McGraw-Hill
A Division of The McGraw-Hill Companies

1 2 3 4 5 6 7 8 9 0 DOC/DOC 9 0 9 8 7 6 5 4 3 2 1 0 9

ISBN 0-07-134652-X

The sponsoring editor for this book was Barbara Gilson, the editing supervisor was Donna Muscatello, and the production supervisor was Tina Cameron. It was set in Stone Serif by PRD Group.

This book is printed on recycled, acid-free paper containing a minimum of 50% recycled de-inked fiber.

Contents

Introduction

Solving word problems in geometry is a challenge, but it can also be fun when you know how. Not only do you have to read and understand word problems and the basics of solving equations, but you also have to know the specialized vocabulary and symbols of geometry. Some words, such as *hypotenuse, isosceles,* and *secant,* are unique to geometry, while others such as *plane, line,* and *construction* are everyday words that take on new meanings.

You also have to know and understand the symbols, and "translate" these symbols into words that allow you to write or think of geometry statements in everyday language. Doing so will help you to "read" a diagram and to draw a diagram from given information without reading more into the diagram than actually exists.

Understanding and working geometry word problems takes practice. The more types of word problems that you do, the better you will become in quickly deciding what the problem asks and reaching a solution.

In this book, you will find a step-by-step approach that gives you solutions to measure your increasing ability to work geometry word problems. Once you have mastered the basics for solving geometry word problems, you will be able to easily apply these principles to even the most challenging advanced problems.

Chapter 1

Points, Lines, Planes, and Angles

Points, lines, and planes are part of your everyday life. When you watch a color television screen or the computer monitor, you are actually viewing a picture that is made up of hundreds of thousands of dots. Because the dots are so small and so close together, your eyes see a complete picture, not the individual dots. Each dot on the screen is similar to a point.

The *point* is the simplest figure that you will study in geometry. Although it doesn't have any size, a point is usually represented by a dot that has size, and it is usually named by a capital letter. All geometric figures consist of points.

One familiar geometric figure is a *line,* a series of connected points that extends in two directions without ending. The picture of a line has some thickness, but the line itself has no thickness. A line is referred to with a single lowercase letter, such as line g, if no points on the line are known. If you know that a line contains points A and B, then you call it line $\overleftrightarrow{AB}$ or simply $\overleftrightarrow{AB}$.

A *plane* is similar to a floor, wall, or tabletop. Unlike these objects, however, a plane extends without ending and has no thickness. Although a plane has no edges, it is usually pictured as a four-sided figure and labeled with a capital letter, plane R. You can think of the ceiling and floor of a room as parts of *horizontal* planes, and the walls as parts of *vertical* planes.

Point, line, and plane are accepted as intuitive ideas in geometry that do not require defining, but they are used in defining other terms in geometry.

Definitions to Know

Acute angle. An angle with measures between 0 and 90°.

Adjacent angles (adj. ∡s). Two angles in a plane that have a common vertex and a common side but no common interior points.

Angle (∡). The figure formed by two rays that have the same endpoint. The rays are called the *sides* of the angle and their common endpoint is the *vertex* of the angle.

Bisector of an angle. The ray that divides the angle into two congruent adjacent angles.

Bisector of a segment. A line, segment, ray, or plane that intersects a line segment at its midpoint.

Collinear points. Points all in one line.

Congruent. Refers to objects that have the same size and shape.

Congruent angles. Angles that have equal measures. To indicate that two angles are congruent, write $m\measuredangle F = m\measuredangle G$ or $m\measuredangle F \cong m\measuredangle G$ ($m\measuredangle$ = measure of angle).

Congruent segments. Line segments that have equal length. To indicate that line segments AB and CD are congruent, write $\overline{AB} \cong \overline{CD}$ or $\overline{AB} = \overline{CD}$.

Coplanar points. Points all in one plane.

Intersection of two figures. The set of points that are in both figures.

Length of a segment. The distance between its endpoints, denoted as $\overline{AB}$. Length must be a positive number.

Midpoint of a segment. The point that divides a line segment into two congruent segments.

Obtuse angle. An angle with a measure between 90 and 180°.

Postulate. In geometry, a statement that is accepted without proof.

Ray of a line. Denoted $\overrightarrow{AD}$, consists of a line segment AD and all other points P, such that D is between A and P. The endpoint of $\overrightarrow{AD}$ is A, the point named first.

Right angle. An angle with a measure of 90°.

Segment of a line. Consists of two points on a line and all points that are between them. It is denoted $\overline{AD}$, in which A and D are the endpoints of the segment.

Space. The set of all points.
Straight angle. An angle with a measure of 180°.
Theorem. In geometry, an important statement that can be proved.

Relevant Postulates and Theorems

Postulate 1 (Ruler Postulate)

1. The points on a line can be paired with the real numbers in such a way that any two points can have coordinates 0 and 1.
2. Once a coordinate system has been chosen in this way, the distance between any two points equals the absolute value of the difference of their coordinates.

Postulate 2 (Segment Addition Postulate)

If B is between A and C, then $\overline{AB} + \overline{BC} = \overline{AC}$.

Postulate 3 (Protractor Postulate)

On $\overleftrightarrow{AB}$ in a given plane, choose any point O between A and B. Consider $\overrightarrow{OA}$ and $\overrightarrow{OB}$ and all the rays that can be drawn from O on one side of $\overleftrightarrow{AB}$. These rays can be paired with the real numbers from 0 to 180 in such a way that

(a) $\overrightarrow{OA}$ is paired with 0, and $\overrightarrow{OB}$ with 180.
(b) If $\overrightarrow{OP}$ is paired with x, and $\overrightarrow{OQ}$ with y, then $m\measuredangle POQ = |x - y|$.

Postulate 4 (Angle Addition Postulate)

If point B lies in the interior of $\measuredangle AOC$, then $m\measuredangle AOB + m\measuredangle BOC = m\measuredangle AOC$. If $\measuredangle AOC$ is a straight angle and B is any point not on $\overleftrightarrow{AC}$, then $m\measuredangle AOB + m\measuredangle BOC = 180°$.

Postulate 5

A line contains at least two points; a plane contains at least three points not all in one line; space contains at least four points not all in one plane.

Postulate 6

Through any two points there is exactly one line.

Postulate 7

Through any three points there is at least one plane, and through any three noncollinear points there is exactly one plane.

Postulate 8

If two points are in a plane, then the line that contains the points is in that plane.

Postulate 9

If two planes intersect, then their intersection is a line.

Theorem 1

If two lines intersect, then they intersect in exactly one point.

Theorem 2

Through a line and a point not in the line there is exactly one plane.

Theorem 3

If two lines intersect, then exactly one plane contains the lines.

EXAMPLE 1

In the diagram, $\overline{PR} \cong \overline{RT}$, S is the midpoint of $\overline{RT}$, $\overline{QR} = 4$, $\overline{ST} = 5$. Find:

(*a*) The value of $\overline{RS}$
(*b*) The value of $\overline{RT}$
(*c*) The value of $\overline{PR}$
(*d*) The value of $\overline{PQ}$

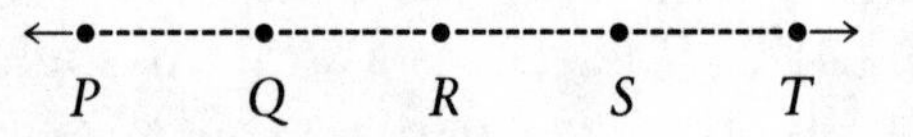

Solution

(*a*) If S is the midpoint of $\overline{RT}$, then the length of $\overline{RS}$ is the same as the length of segment $\overline{ST}$. We are told that the length of segment $\overline{ST}$ is 5.

$$\overline{RS} = \overline{ST} \text{ and } \overline{ST} = 5$$

$$\text{If } \overline{ST} = 5, \text{ then } \overline{RS} = 5$$

(*b*) The length of segment $\overline{RT}$ is the sum of the two segments $\overline{RS}$ and $\overline{ST}$. We already know that $\overline{RS} = 5$ and $\overline{ST} = 5$.

$$\overline{RT} = \overline{RS} + \overline{ST} = 5 + 5 = 10$$

(*c*) The problem tells us that the length of line segment *PR* is equal to the length of line segment *RT*. From solving part (*b*), we know that $\overline{RT} = 10$.

$$\overline{PR} \cong \overline{RT}$$
$$\overline{RT} = 10$$
$$\overline{PR} = 10$$

(*d*) The problem tells us that the length of line segment *QR* is 4. We know that $\overline{PR}$ is the sum of line segments *PQ* and *QR*. From solving part (*c*), we know that the length of $\overline{PR}$ is 10.

$$\overline{PQ} + \overline{QR} = \overline{PR}$$
$$\overline{PQ} + 4 = 10$$
$$\overline{PQ} = 6$$

EXAMPLE 2

M is a point on line segment *LN* in which $\overline{LM} = x + 8$, $\overline{MN} = x$, and $\overline{LN} = 32$. Find:

(*a*) The value of *x*
(*b*) $\overline{LM}$

$$\underset{L}{\bullet}\text{- - - - - } x + 8 \text{ - - - - -}\underset{M}{\bullet}\text{- - - - } x \text{ - - - -}\underset{N}{\bullet}$$

Solution

(*a*)
$$\overline{LM} + \overline{MN} = \overline{LN}$$
$$(x + 8) + x = 32$$
$$2x = 24$$
$$x = 12$$

(*b*) $\overline{LM} = x + 8 = 12 + 8 = 20$

EXAMPLE 3

In Fig. 1-1, identify the measures of the remaining angles when the measure of $\measuredangle 1$ is 93°.

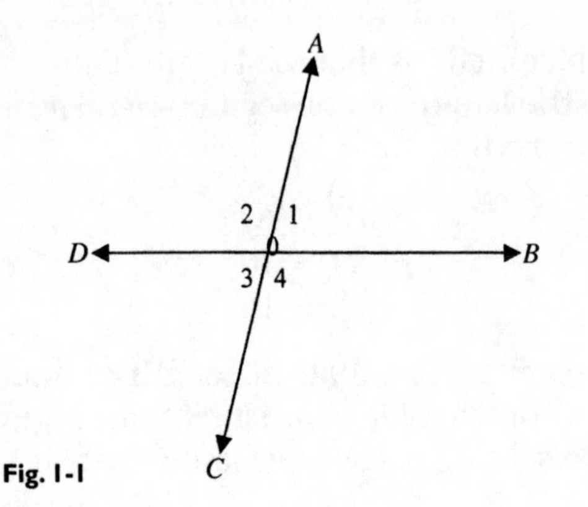

Fig. 1-1

Solution

$$m\measuredangle 2 = m\measuredangle AOD \text{ (a straight angle of measure 180°)}$$
$$m\measuredangle 1 + m\measuredangle 2 = 180$$
$$93 + m\measuredangle 2 = 180$$
$$m\measuredangle 2 = 87°$$

$$m\measuredangle 1 + m\measuredangle 4 = m\measuredangle AOC \text{ (a straight angle of measure 180°)}$$
$$m\measuredangle 1 + m\measuredangle 4 = 180$$
$$93 + m\measuredangle 4 = 180$$
$$m\measuredangle 4 = 87°$$

$$m\measuredangle 2 + m\measuredangle 3 = m\measuredangle AOC \text{ (a straight angle of measure 180°)}$$
$$m\measuredangle 2 + m\measuredangle 3 = 180$$
$$87 + m\measuredangle 3 = 180$$
$$m\measuredangle 3 = 93°$$

EXAMPLE 4

Express the measure of all angles in Fig. 1-2 in terms of t when $m\measuredangle 1 = t$ and the intersection of two lines is O.

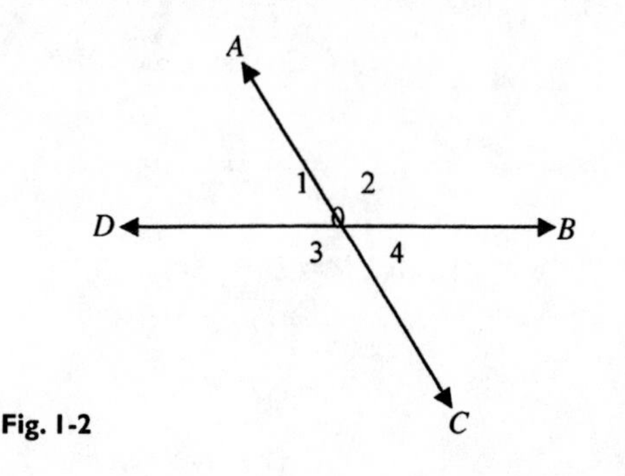

Fig. 1-2

Solution

$\measuredangle AOC$ is a straight angle that has a measure of 180°. $\measuredangle DOB$ is a straight angle that has a measure of 180°.

$$m\measuredangle 1 + m\measuredangle 4 = m\measuredangle AOC$$
$$t + m\measuredangle 4 = 180$$
$$m\measuredangle 4 = 180° - t$$

$$m\measuredangle 1 + m\measuredangle 2 = m\measuredangle DOB$$
$$t + m\measuredangle 2 = 180$$
$$m\measuredangle 2 = 180° - t$$

$$m\measuredangle 2 + m\measuredangle 3 = m\measuredangle AOC$$
$$(180 - t) + m\measuredangle 3 = 180$$
$$m\measuredangle 3 = 180 - (180 - t)$$
$$m\measuredangle 3 = t$$

EXAMPLE 5

$\measuredangle AOB$ and $\measuredangle BOC$ are adjacent angles which form an angle of 98°. The measure of $\measuredangle APB$ is 36° more than the measure of $\measuredangle BOC$. Find $m\measuredangle AOB$.

Solution

Let $m\measuredangle AOB$ equal $b + 36$ and $m\measuredangle BOC$ equal b. If the sum of the two angles is 98°, then the following is true:

$$m\measuredangle AOB + m\measuredangle BOC = 98°$$
$$(b + 36) + b = 98$$

$$2b + 36 = 98$$
$$2b = 62$$
$$b = 31°$$

$$m\measuredangle AOB = b + 36$$
$$= 31 + 36$$
$$= 67°$$

Supplementary Lines and Angles Problems

R ●------------● N ------------------● C

1. When $\overline{RN} = 7$, $\overline{NC} = 3x + 5$, and $\overline{RC} = 18$, what is the value of x?
2. When point N bisects $\overline{RC}$, $\overline{RN} = x + 7$, and $\overline{RC} = 28$, what is the value of x?
3. When $\overline{RN} = x$, $\overline{NC} = x - 7$, and $\overline{RC} = 29$, what is the value of x?
4. Find the length of a line segment AD if $\overline{AC}$ is 8 and C is the midpoint of $\overline{AD}$.
5. Find the measure of two angles that are adjacent and form an angle of 160°. The measure of the larger angle is 20° greater than the measure of the smaller angle.
6. Find the measure of two adjacent angles that form an angle measuring 85°. The difference between the two angles is 19°.
7. Find the measures of two angles that are adjacent and form an angle measuring 150° if one angle is 36° more than twice the measure of the other angle.
8. Find the measures of the angles formed by the hands of a clock at 4 P.M.

Solutions to Supplementary Lines and Angles Problems

1. You will find that solving problems that involve coordinates on a line segment is easier if you place the values given on the line, so that you have a visual image of the problem.

R ●----- 7 -----● N ----- $x + 7$ -----● C

$$\overline{RN} + \overline{NC} = \overline{RC}$$
$$7 + (3x + 5) = 18$$
$$3x + 12 = 18$$
$$3x = 6$$
$$x = 2$$

2. Draw the line segment and, once again, identify the lengths of line segments with the values provided in the problem. *Tip:* Even if the diagram does not appear to be bisected by a given point, follow the directions of the problem and use the coordinates as given.

$x + 7$ $\quad$ $x + 7$

R $\quad$ N $\quad$ C

$$(x + 7) + (x + 7) = 28$$
$$2x + 14 = 28$$
$$2x = 14$$
$$x = 7$$

3. Draw the line segment again and identify the lengths of line segments with the values provided in the problem.

x $\quad$ $x - 7$

R $\quad$ N $\quad$ C

$$x + (x - 7) = 29$$
$$2x - 7 = 29$$
$$2x = 36$$
$$x = 18$$

4. If C is the midpoint of $\overline{AD}$, the C must be a point located on the line segment AD. The length of $\overline{AC}$ equals the length of $\overline{AD}$.

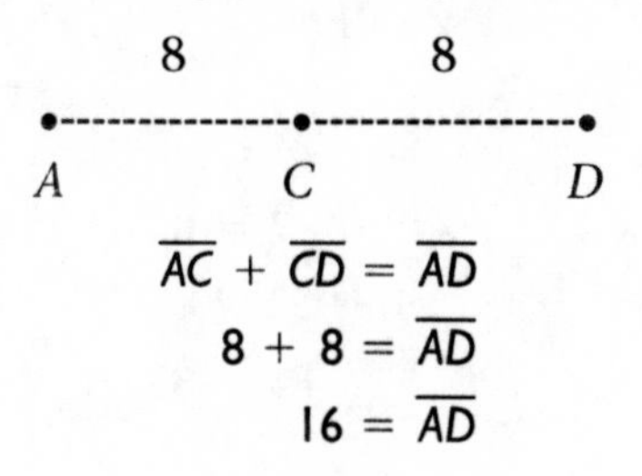

$$\overline{AC} + \overline{CD} = \overline{AD}$$
$$8 + 8 = \overline{AD}$$
$$16 = \overline{AD}$$

5. To determine the individual measures of two adjacent angles that form a third, larger angle, first create equations for each of the smaller angles.

$$\measuredangle 1 = x$$
$$\measuredangle 2 = x + 20°$$

When added, the two angles form a third angle that measures 160°.

$$\measuredangle 1 + \measuredangle 2 = 160$$
$$x + (x + 20) = 160$$
$$2x + 20 = 160$$
$$2x = 140$$
$$x = \measuredangle 1 = 70°$$
$$x + 20 = \measuredangle 2 = 90°$$

Check:

$$\measuredangle 1 + \measuredangle 2 = 160$$
$$70 + 90 = 160$$
$$160 = 160$$

6. To determine the individual measures of two adjacent angles that form a third, larger angle, first create equations for each of the small angles. Let the measure of the smaller angle $\measuredangle 1$ equal x and the measure of the larger angle $\measuredangle 2$ equal $x + 19°$. Their sum is 85°.

$$\measuredangle 1 + \measuredangle 2 = 85$$
$$x + (x + 19) = 85$$
$$2x + 19 = 85$$
$$2x = 66$$
$$x = \measuredangle 1 = 33°$$
$$x + 19 = \measuredangle 2 = 52°$$

Check:

$$\measuredangle 1 + \measuredangle 2 = 85$$
$$33 + 52 = 85$$
$$85 = 85$$

7. To determine the individual measures of two adjacent angles that form a third, larger angle, first create equations for each of the small angles. Let the measure of the smaller angle $\measuredangle 1$ equal x and the measure of the second angle $\measuredangle 2$, which is twice the size of the first angle plus 36°, equal $2x + 36$.

$$\begin{aligned} x + (2x + 36) &= 150 \\ 3x + 36 &= 150 \\ 3x &= 114 \\ x &= \measuredangle 1 = 38^\circ \\ 2x + 36 &= \measuredangle 2 = 112^\circ \end{aligned}$$

Check:

$$\begin{aligned} \measuredangle 1 + \measuredangle 2 &= 150 \\ 38 + 112 &= 50 \\ 150 &= 150 \end{aligned}$$

8. The measure of the circular face of a clock equals 360°. The angle formed by placing the hands of the clock at 12 and 4 forms an angle $\measuredangle 1$ that is $\frac{4}{12}$ of the number of units (12) that make up the clock face. Set up the equation as follows:

$$\measuredangle 1 = \frac{4}{12}(360) = 4\ (30) = 120^\circ$$

Chapter 2

Deductive Reasoning

Reasoning skills are very important in helping you to make decisions in everyday life. You observe situations, people, and actions and try to explain their connections. Sometimes your assumptions are correct, while at other times you are completely wrong. This method of *direct observation* is also useful in geometry.

Inductive reasoning, the method of direct observation to study a series of specific examples to discover general relationships, is used in geometry, but the results are not always accurate. Why not? You may examine all available examples and come to a general conclusion about the characteristics of a figure. Later, you may use the same inductive reasoning on another figure and the results will show the general conclusion to be false. The possibility of a *counterexample,* an example which shows the general conclusion to be false, is enough to make the results of inductive reasoning only *probable,* not definite.

Consider this example. You might examine several *right triangles,* those that have a 90° angle, and find that in all your examples one of two remaining angles is always larger than the other. Based on this observation, you might conclude that this is true of all right triangles. When someone shows you a right triangle in which the two remaining angles are each 45°, this counterexample shows that your generalization is incorrect.

Deductive reasoning is a more accurate approach, and it is used to prove statements through the use of given information, theorems, definitions, and postulates to arrive at conclusions. In its most simple form, deductive reasoning takes the form of an *if-then statement.* You use these every day in dealing with other people. For example, you might tell a friend the following: "If you can help me to complete my project, then I will have the time to proofread your paper." The *condition* is set in the "if" part of the statement and the *conclusion,* or possible result, appears in the "then" part of the statement. To make this statement "true," however, the condition has to be true. If your friend does not help you to complete the project, then the conclusion cannot occur.

How does this relate to geometry and, more specifically, to deductive reasoning? Deductive reasoning depends upon such conditional statements. In geometry, the condition is given another name, *hypothesis*. We might make the following statement:

If $\overline{RT}$ is equal to $\overline{RS} + \overline{ST}$, then $\overline{RS}$ is shorter in length than $\overline{RT}$.

The above statement can be rephrased in this basic form:

If h, then c.

↑ ↑

h: hypothesis c: conclusion

Be careful in choosing the information on which you base the hypothesis, so that you will reach a *true* and *valid* conclusion. You only need to find one counterexample, an example for which the hypothesis is true and the conclusion is false, to disprove a statement.

For a statement to be a *definition,* a given or an accepted truth, both the conditional statement and its *converse,* formed by interchanging the hypothesis and the conclusion, must be true. The following statement and its converse are both true.

Statement: If $\measuredangle M$ is a straight angle, then the measure of $\measuredangle M$ is 180°.

Converse: If the measure of $\measuredangle M$ is 180°, then $\measuredangle M$ is a straight angle.

Conditional statements in deductive reasoning are not always phrased in the if-then form, so learn the other ways in which this same concept may be stated. Despite their different language, all the conditional statements below mean the same.

Statement Form	**Example**
If *h*, then *c*.	If $\measuredangle AOC$ is a right angle, then $\measuredangle AOC = 90°$.
h implies *c*.	$\measuredangle AOC$ is a right angle implies $\measuredangle AOC = 90°$.
h only if *c*.	$\measuredangle AOC$ is a right angle only if $\measuredangle AOC = 90°$.
c if *h*.	$\measuredangle AOC = 90°$ if $\measuredangle AOC$ is a right angle.

In a definition both the hypothesis and converse are true. Every definition can be written as a *biconditional* statement, one that contains the words "if and only if," as the following statement and example show.

Statement: *h* if and only if *c*.

Example: *Definition.* A straight angle is one that measures 180°.

Biconditional. An angle is a straight angle if and only if it measures 180°.

Definitions to Know

Complementary angles. Two angles whose measures add up to 90°, making each angle the *complement* of the other.

Deductive reasoning. A method of proving statements by using accepted postulates, definitions, theorems, and given information.

Perpendicular lines. Lines that intersect to form right (90°) angles.

Proofs. Examples of deductive reasoning used to prove statements in geometry through the use of reasons such as given information, definitions, postulates (including properties from algebra), and theorems. A well-reasoned proof includes four sections: (1) a lettered figure that illustrates the given information; (2) a list, in terms of the lettered figure, of the information that is given (the hypothesis); (3) a list, in terms of the diagram, of what you are to prove (the conclusion); and (4) a series of logical arguments used in demonstrating the proof, phrased as a series of steps that each contains a statement and a reason and which lead from the given information to the statement that is to be proved.

Properties. Statements in algebra that are accepted as true and treated as postulates in geometry.

Supplementary angles. Two angles whose measures add up to 180°, making each angle the *supplement* of the other.

Vertical angles. Two angles formed by intersecting lines, where the sides of one angle are opposite rays to the sides of the other angle. In Fig. 2-1, $\measuredangle 1$ and $\measuredangle 2$ are vertical angles, as are $\measuredangle 3$ and $\measuredangle 4$.

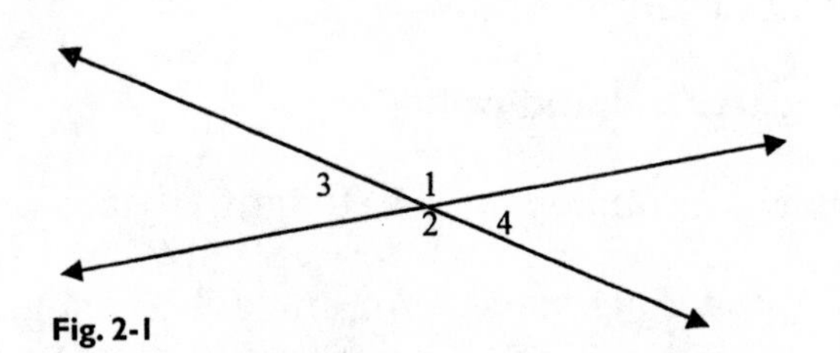

Fig. 2-1

Relevant Postulates and Theorems

In geometry, we treat as postulates the properties of equality that students learn in algebra.

Postulate 1 (Addition Property)

If $a = b$ and $c = d$, then $a + c = b + d$.

or

If $\overline{GH} = \overline{IJ}$ and $\overline{KL} = \overline{MN}$, then $\overline{GH} + \overline{KL} = \overline{IJ} + \overline{MN}$.

and

If $\measuredangle ABC = \measuredangle DEF$ and $\measuredangle GHI = \measuredangle JKL$, then $\measuredangle ABC + \measuredangle GHI = \measuredangle DEF + \measuredangle JKL$.

Postulate 2 (Subtraction Property)

If $a = b$ and $c = d$, then $a - c = b - d$.

or

If $\overline{GH} = \overline{IJ}$ and $\overline{KL} = \overline{MN}$, then $\overline{GH} - \overline{KL} = \overline{IJ} - \overline{MN}$.

and

If $\measuredangle ABC = \measuredangle DEF$ and $\measuredangle GHI = \measuredangle JKL$, then $\measuredangle ABC - \measuredangle GHI = \measuredangle DEF - \measuredangle JKL$.

Postulate 3 (Multiplication Property)

If $a = b$, then $ca = cb$.

or

If $\overline{GH} = \overline{IJ}$, then $(a)\overline{GH} = (a)\overline{IJ}$.

and

If $\measuredangle ABC = \measuredangle DEF$, then $(a)(\measuredangle ABC) = (a)(\measuredangle DEF)$.

Postulate 4 (Division Property)

If $a = b$ and $c \neq 0$, then $a/c = b/c$.

or

If $\overline{GH} = \overline{IJ}$, then $\overline{GH}/a = \overline{IJ}/a$.

and

If $\measuredangle ABC = \measuredangle DEF$, then $\measuredangle ABC \div a = \measuredangle DEF \div a$.

Postulate 5 (Substitution Property)

If $a = b$, then either a or b may be substituted for the other in any equation or inequality.

or

If $\overline{AB} + \overline{CD}$, then either $\overline{AB}$ or $\overline{CD}$ may be substituted for the other in any equation or inequality.

and

If $\measuredangle ABC = \measuredangle DEF$, then either $\measuredangle ABC$ or $\measuredangle DEF$ may be substituted for the other in any equation or inequality.

Postulate 6 (Reflexive Property)

A quantity is equal to itself:

$$a = a$$

or

$$\overline{AB} \cong \overline{AB}$$

and

$$\measuredangle ABC \cong \measuredangle ABC$$

Postulate 7 (Symmetric Property)

If $a = b$, then $b = a$.

or

If $\overline{GH} \cong \overline{LM}$, then $\overline{LM} \cong \overline{GH}$.

and

If $\measuredangle ABC \cong \measuredangle DEF$, then $\measuredangle DEF \cong \measuredangle ABC$.

Postulate 8 (Transitive Property)

If $a = b$ and $b = c$, then $a = c$.

or

If $\overline{AB} \cong \overline{DE}$ and $\overline{DE} \cong \overline{FG}$, then $\overline{AB} \cong \overline{FG}$.

and

If $\measuredangle ABC \cong \measuredangle DEF$ and $\measuredangle DEF \cong \measuredangle GHI$, then $\measuredangle ABC \cong \measuredangle GHI$.

Theorem 1

If M is the midpoint of $\overline{XY}$, then $\overline{XM} = \frac{1}{2}\overline{XY}$ and $\overline{MY} = \frac{1}{2}\overline{XY}$.

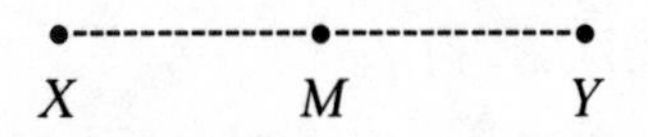

Theorem 2

If $\overrightarrow{XZ}$ is the bisector of $\measuredangle WXY$, then $m\measuredangle WXZ = \frac{1}{2}m\measuredangle WXY$ and $m\measuredangle ZXY = \frac{1}{2}m\measuredangle WXY$. (See Fig. 2-2.)

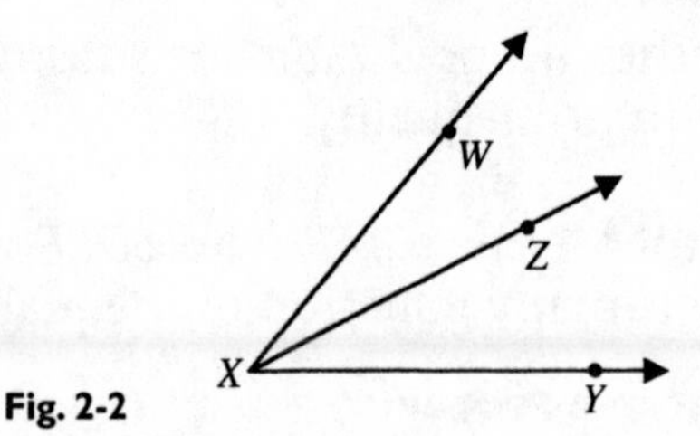

Fig. 2-2

Theorem 3

If two angles are vertical angles, then they are congruent.

Theorem 4

If two lines are perpendicular, then they form congruent adjacent angles, as in Fig. 2-3.

Theorem 5

If two lines form congruent adjacent angles, then the lines are perpendicular, as in Fig. 2-3.

Theorem 6

If the exterior sides of two adjacent acute angles are perpendicular, then the adjacent angles are complementary, as in Fig. 2-3.

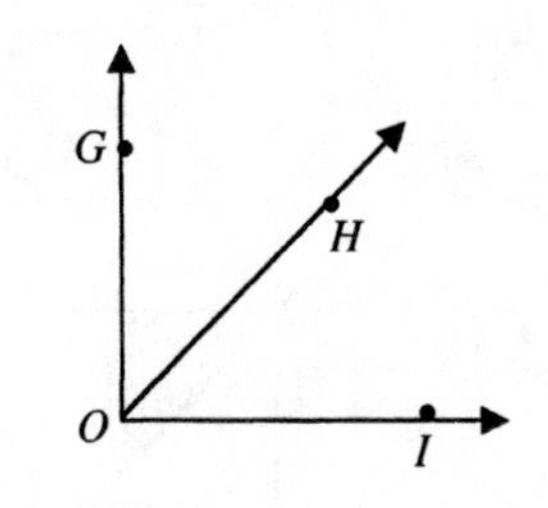

Fig. 2-3

Theorem 7

If two angles are supplements of congruent angles (or of the same angle), then the two angles are congruent.

Theorem 8

If two angles are complements of congruent angles (or of the same angle), then the two angles are congruent.

EXAMPLE 1

Use the following conditional statement to answer the questions below: If $m\measuredangle AOB = 100°$, then $\measuredangle AOB$ is obtuse.

(*a*) What is the hypothesis?
(*b*) What is the conclusion?
(*c*) What is the converse?
(*d*) Is the conclusion valid?

Solution

(*a*) $m\measuredangle AOB = 100°$.
(*b*) $\measuredangle AOB$ is obtuse.
(*c*) If $\measuredangle AOB$ is obtuse, then $m\measuredangle AOB = 100°$.

(*d*) The conclusion is valid because the definition of an obtuse angle is one with a measure between 90 and 180°.

EXAMPLE 2

Name the postulate, definition, or theorem that justifies each statement about Fig. 2-4.

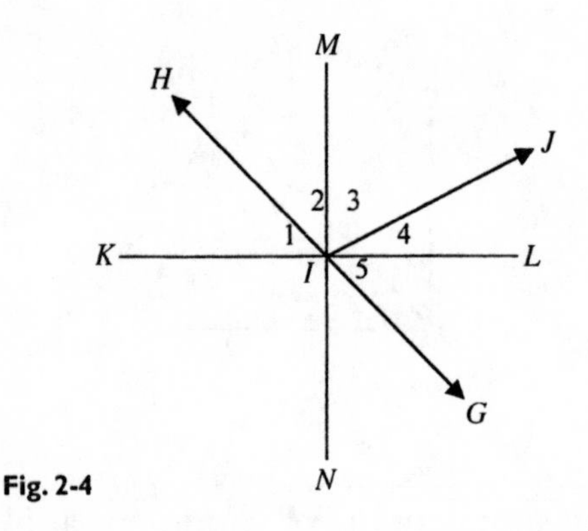

Fig. 2-4

(*a*) $\overline{MJ} + \overline{JN} = \overline{MN}$
(*b*) $\measuredangle 1 \cong \measuredangle 5$
(*c*) If $\overline{IJ}$ bisects $\measuredangle MIL$, then $\measuredangle 3 \cong \measuredangle 4$.
(*d*) If $\overline{MI} \perp \overline{KL}$, then $m\measuredangle MIL = 90°$.
(*e*) If $m\measuredangle 5 + m\measuredangle 6 = 90°$, then $\measuredangle 5$ and $\measuredangle 6$ are complements.
(*f*) If $\overline{MN} \perp \overline{KL}$, then $\measuredangle MIL \cong \measuredangle KIN$.
(*g*) If $\measuredangle JIH$ is a right angle, then $\overrightarrow{JI} \perp \overrightarrow{IH}$.

Solution

(*a*) Segment Addition Postulate (see Chap. 1).
(*b*) Theorem 3: Vertical angles are congruent.
(*c*) Theorem 2: Definition of angle bisection.
(*d*) Definition of perpendicular lines.
(*e*) Definition of complementary angles.
(*f*) Theorem 4: If two lines are perpendicular, then they form congruent adjacent angles.
(*g*) Definition of perpendicular lines.

EXAMPLE 3

Write a two-column proof for Fig. 2-5.

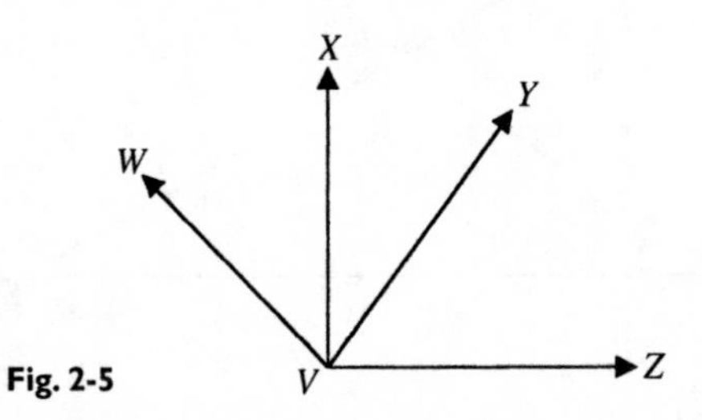

Fig. 2-5

Given: $\measuredangle XVY$ is complementary to $\measuredangle YVZ$
$\measuredangle WVX$ is complementary to $\measuredangle XVY$

Prove: $\measuredangle WVX \cong \measuredangle YVZ$

Statements	Reasons
1. $\measuredangle YVZ$ is complementary to $\measuredangle XVY$	1. Given
2. $m\measuredangle XVY + m\measuredangle YVZ = 90°$	2. Definition of complementary angles
3. $\measuredangle WVX$ is complementary to $\measuredangle XVY$	3. Given
4. $m\measuredangle WVX + m\measuredangle XVY = 90°$	4. Definition of complementary angles
5. $m\measuredangle YVZ + m\measuredangle XVY = m\measuredangle WVX + m\measuredangle XVY$	5. Postulate 8: Transitive postulate of equality
6. $m\measuredangle XVY = m\measuredangle XVY$	6. Postulate 6: Reflexive property of equality
7. $m\measuredangle YVZ = m\measuredangle WVX$	7. Postulate 2: Subtraction property of equality
8. $\measuredangle YVZ \cong m\measuredangle WVX$	8. Definition of congruent angles: if the measures of two angles are equal, then the angles are congruent.

Supplementary Deductive Reasoning Problems

Refer to Fig. 2-6 for Probs. 1 through 5.

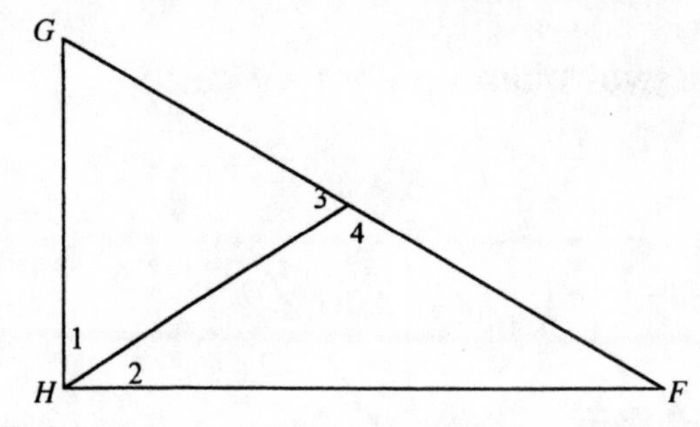

Fig. 2-6

1. If $\overline{GH} \perp \overline{GF}$, which two angles in the figure are complementary?
2. Which two angles in the figure are supplementary?
3. If $m\measuredangle 2 = 38°$, what is the measure of $\measuredangle 1$?
4. If $m\measuredangle 3 = 62°$, what is the measure of $\measuredangle 4$?
5. If $m\measuredangle 2 = 38°$ and $m\measuredangle 3 = 62°$, what is the measure of $\measuredangle HFG$?
6. Create a proof for Fig. 2-7.

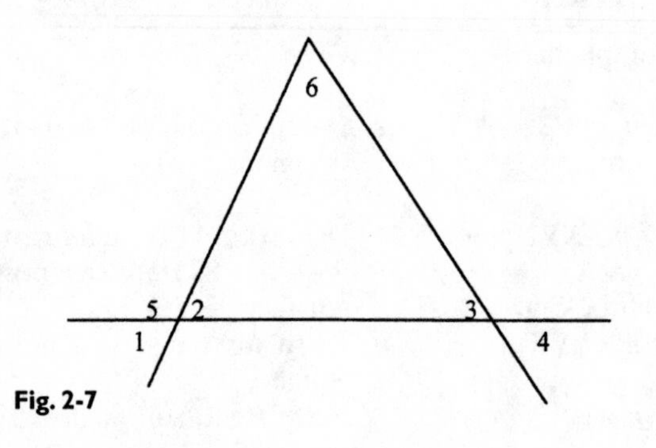

Fig. 2-7

Given: $\measuredangle 2 \cong \measuredangle 3$

Prove: $\measuredangle 1 \cong \measuredangle 4$

Refer to Fig. 2-7 for Probs. 7 through 10.

7. If $m\measuredangle 4 = 53°$, what is the measure of $\measuredangle 3$?
8. If $m\measuredangle 1 = 72°$, what is the measure of $\measuredangle 5$?
9. If $m\measuredangle 1 = 72°$, what is the measure of $\measuredangle 2$?
10. If $m\measuredangle 1 = 72°$ and $m\measuredangle 4 = 63°$, what is the measure of $\measuredangle 6$?

Solutions to Supplementary Deductive Reasoning Problems

1. If $\overline{GH} \perp \overline{GF}$, then $\measuredangle GHF$ is a right angle. Because $\measuredangle 1$ ane $\measuredangle 2$ are formed by the division of the right angle, they are complementary angles by definition.
2. The intersection of a line with straight $\measuredangle GF$ forms $\measuredangle 3$ and $\measuredangle 4$, making them supplementary angles—two angles whose sum is 180°—by definition.
3. Because $\measuredangle 1$ and $\measuredangle 2$ are complementary, their sum is 90°. If the measure of $\measuredangle 2 = 38°$, then you must subtract the measure of $\measuredangle 2$ from 90° to obtain the measure of $\measuredangle 1$.

$$m\measuredangle 1 = 90° - m\measuredangle 2 = 90° - 38° = 52°$$

4. Because $\measuredangle 3$ and $\measuredangle 4$ are supplementary, their sum is 180°. If the measure of $\measuredangle 3$ is 62°, then the measure of its supplement $\measuredangle 4$ must be 118°.
5. The measure of $\measuredangle 3$ is 62°, and the measure of its supplement $\measuredangle 4$ is 118°. The sum of the measures of the interior angles of a triangle is 180°. You must add the measures of the two angles that you know and subtract that sum from 180° to obtain the measure of the third angle.

$$180° - (m\measuredangle 2 + m\measuredangle 4) = m\measuredangle HFG$$
$$180° - (38° + 118°) = m\measuredangle HFG$$
$$180° - 156° = 24°$$

6.

Statements	Reasons
1. $\measuredangle 1 \cong \measuredangle 2$	1. Vertical angles are congruent.
2. $\measuredangle 2 \cong \measuredangle 3$	2. Given
3. $\measuredangle 1 \cong \measuredangle 3$	3. Postulate 8: The transitive postulate of equality
4. $\measuredangle 3 \cong \measuredangle 4$	4. Vertical angles are congruent.
5. $\measuredangle 1 \cong \measuredangle 4$	5. Postulate 8: The transitive postulate of equality

7. The measure of $\measuredangle 3$ is 53°, because $\measuredangle 3$ and $\measuredangle 4$ are vertical angles, which by definition are equal and congruent. Thus, if $m\measuredangle 4$ is 53°, then $m\measuredangle 3 = 53°$.

8. The sum of the measures of $\measuredangle 1$ and $\measuredangle 5$ is 180°, because they are supplementary angles. Therefore, the measure of $\measuredangle 5$ is equal to the following:

$$m\measuredangle 5 = 180° - m\measuredangle 1 = 180° - 72° = 108°$$

9. The measure of $\measuredangle 2$ is 53°, because $\measuredangle 1$ and $\measuredangle 2$ are vertical angles, which by definition are equal and congruent. Thus, if $m\measuredangle 1$ is 72°, then $m\measuredangle 2 = 72°$.

10. The sum of the interior angles of a triangle is 180°, so $m\measuredangle 2 + m\measuredangle 3 + m\measuredangle 6 = 180°$. In Probs. 7 and 9, we have determined that $m\measuredangle 2 = 72°$ and $m\measuredangle 3 = 53°$.

$$\begin{aligned} m\measuredangle 2 + m\measuredangle 3 + m\measuredangle 6 &= 180° \\ 72° + 53° + m\measuredangle 6 &= 180° \\ 125° + m\measuredangle 6 &= 180° \\ m\measuredangle 6 &= 55° \end{aligned}$$

Chapter 3

Parallel Lines and Planes

You have identified some of the basic terms and relationships between points, lines, and angles. Now you are ready to identify the different line and plane relationships, as well as to identify the results of the intersections of different types of lines and planes.

Definitions to Know

Refer to Fig. 3-1 to help you to understand the definitions below.

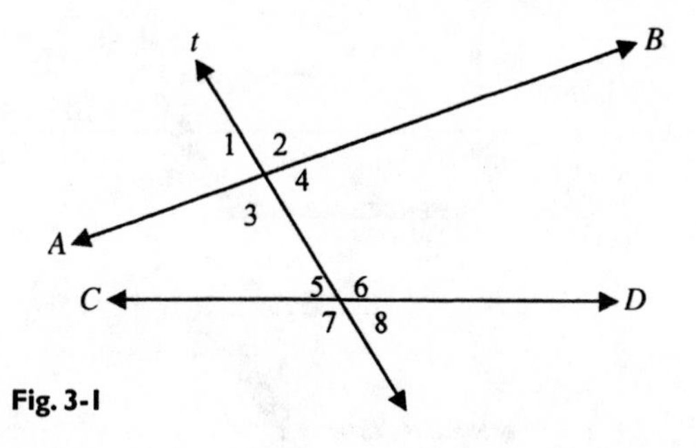

Fig. 3-1

Alternate interior angles. Two nonadjacent interior angles on opposite sides of a transversal. In Fig. 3-1, the following angles are alternate interior angles: ∡4 and ∡5, ∡3 and ∡6.

Coplanar lines. Lines whose points are all in one plane.

Corresponding angles. Two angles that appear in the same position in relation to the two lines cut by a transversal. In Fig. 3-1, the following angles are corresponding angles: $\measuredangle 1$ and $\measuredangle 5$, $\measuredangle 3$ and $\measuredangle 7$, $\measuredangle 2$ and $\measuredangle 6$, and $\measuredangle 4$ and $\measuredangle 8$.

Exterior angles. Angles that are formed outside of the coplanar lines that are cut by a transversal. In Fig. 3-1, the following are exterior angles: $\measuredangle 1$, $\measuredangle 2$, $\measuredangle 7$, and $\measuredangle 8$.

Interior angles. Angles that are formed between the coplanar lines that are cut by a transversal. In Fig. 3-1, the following are interior angles: $\measuredangle 3$, $\measuredangle 4$, $\measuredangle 5$, and $\measuredangle 6$.

Parallel lines. Lines that are coplanar and that do not intersect. The corresponding points on the lines are equidistant from each other. (See Fig. 3-2.)

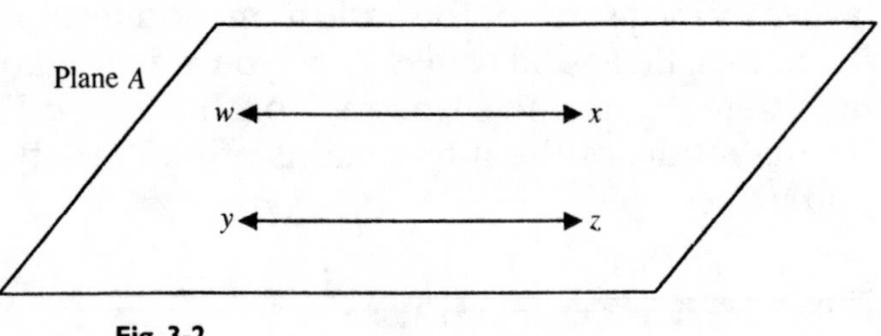

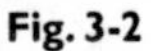
Fig. 3-2

Parallel planes. Planes that do not intersect. (See Fig. 3-3.)

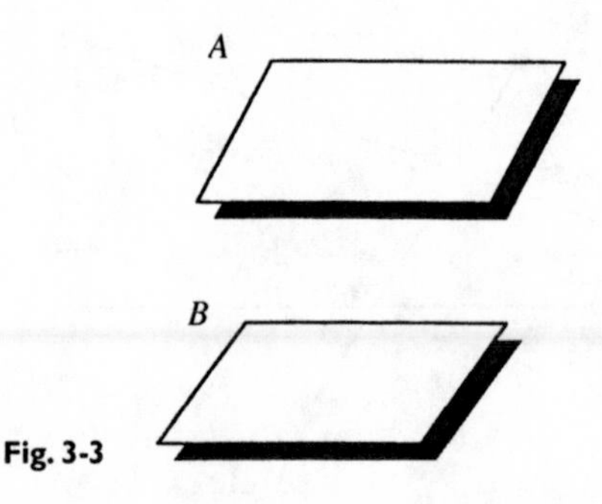

Fig. 3-3

Same-side interior angles. Two interior angles that are on the same side of the transversal and between the coplanar lines. In Fig. 3-1, the following are same-side interior angles: $\measuredangle 3$ and $\measuredangle 5$, $\measuredangle 4$ and $\measuredangle 6$.

Skew lines. Lines that are noncoplanar and, as a result, they are not parallel nor can they intersect. (See Fig. 3-4.)

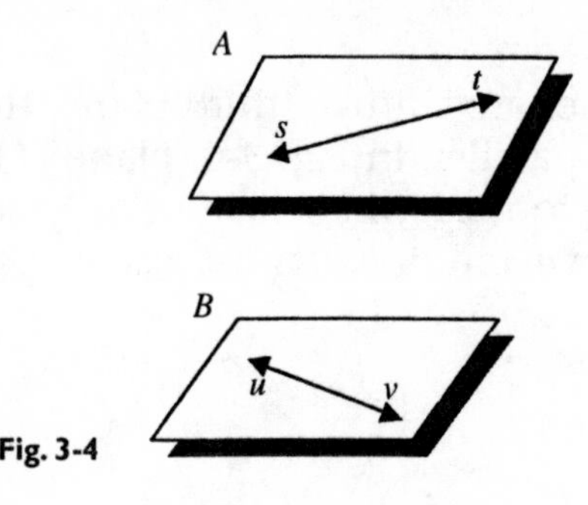

Fig. 3-4

Transversal. A line that intersects two or more parallel or nonparallel coplanar lines. (See Fig. 3-5.)

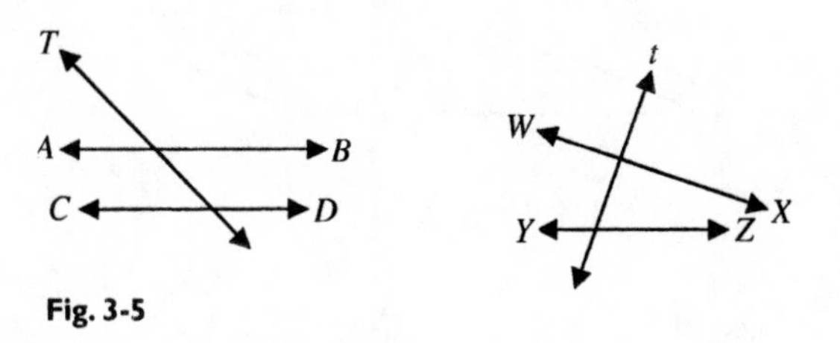

Fig. 3-5

Relevant Postulates and Theorems

Postulate 1

Through a point outside a line, there is exactly one line parallel to the line.

Postulate 2

Through a point outside a line, there is exactly one line perpendicular to the given line.

Postulate 3

Two lines parallel to a third line are parallel to each other.

Postulate 4

If two parallel lines are cut by a transversal, then corresponding angles in the figure that they form are congruent.

Postulate 5

If two lines are cut by a transversal and their corresponding angles are congruent, then the lines are parallel.

Theorem 1

If two parallel planes are cut by a third plane, then the lines of intersection are parallel. In Fig. 3-6, plane D is parallel to plane E. The lines formed when plane F intersects the two parallel planes, line a and line b, must also be parallel.

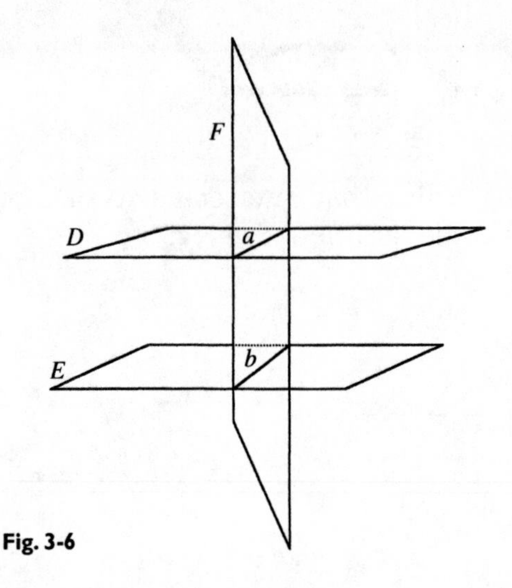

Fig. 3-6

Theorem 2

If two parallel lines are cut by a transversal, then the alternate interior angles formed are congruent. Thus, in Fig. 3-7, $\measuredangle 1 \cong \measuredangle 2$ and $\measuredangle 3 \cong \measuredangle 4$.

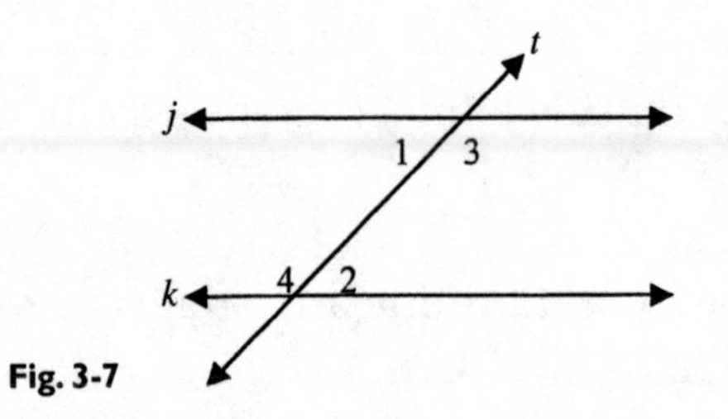

Fig. 3-7

Theorem 3

If two parallel lines are cut by a transversal, then same-side interior angles are supplementary. Thus, in Fig. 3-8, $\measuredangle 1$ and $\measuredangle 2$ are supplementary, and $m\measuredangle 1 + m\measuredangle 2 = 180°$.

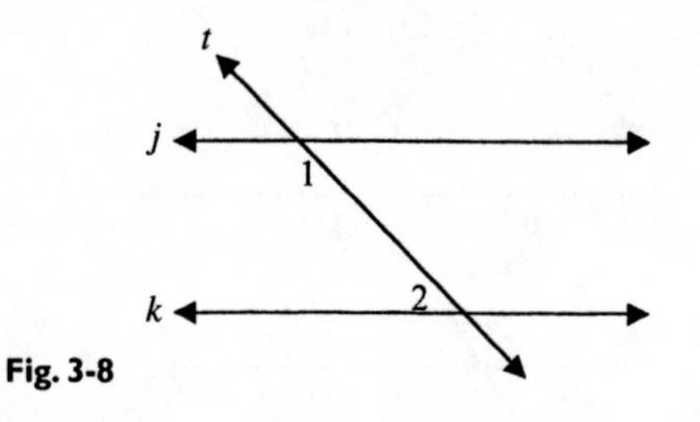

Fig. 3-8

Theorem 4

If a transversal is perpendicular to one of two parallel lines, then it is perpendicular to the other one as well. In Fig. 3-9, $m\measuredangle 1 = m\measuredangle 2 = m\measuredangle 3 = m\measuredangle 4 = 90°$.

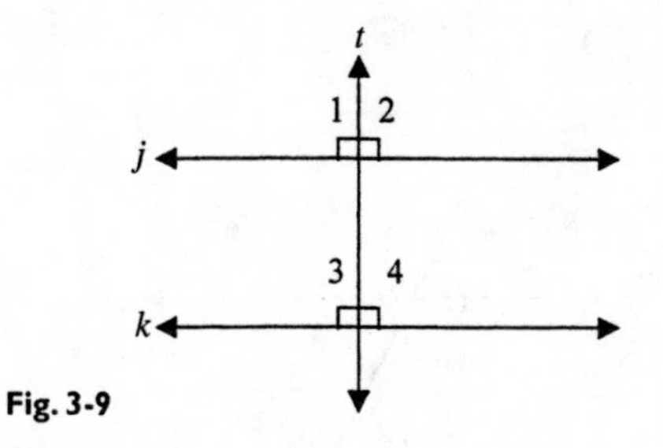

Fig. 3-9

Theorem 5

If two coplanar lines are perpendicular to the same line, then the lines are parallel. In Fig. 3-10, $m\measuredangle 1 = m\measuredangle 2 = 90°$.

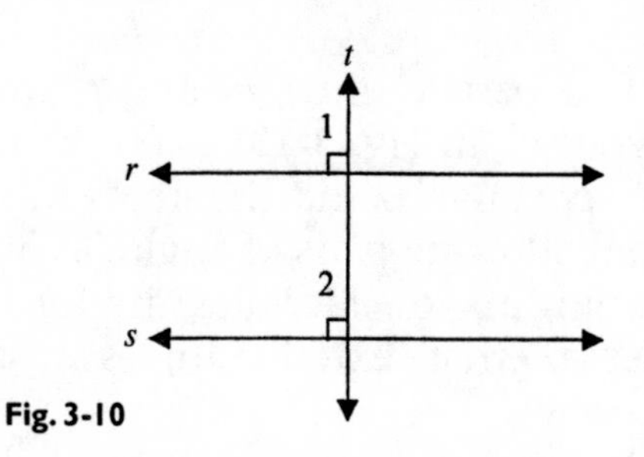

Fig. 3-10

Theorem 6

If two lines are cut by a transversal and alternate interior angles are congruent, then the lines must be parallel. In Fig. 3-11, $\measuredangle 1 \cong \measuredangle 2$.

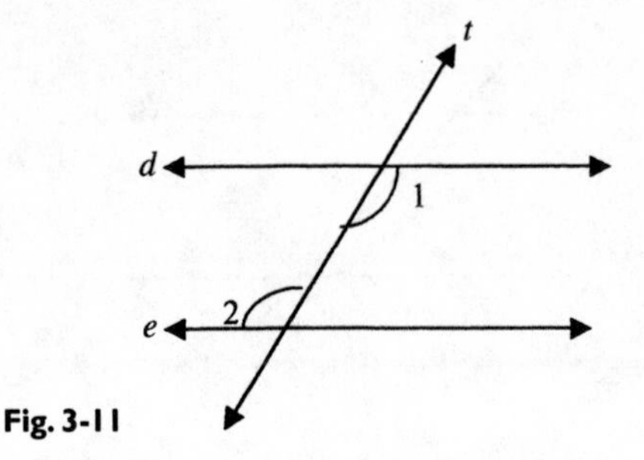

Fig. 3-11

Theorem 7

If two lines are cut by a transversal and same-side interior angles are supplementary, then the lines must be parallel. In Fig. 3-12, $m\measuredangle 1 + m\measuredangle 2 = 180°$.

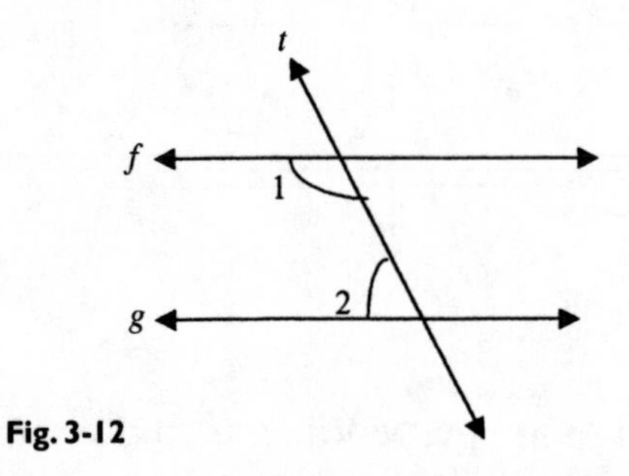

Fig. 3-12

EXAMPLE 1

In a plane, two parallel lines *m* and *p* are cut by the transversal *t*, as you see in Fig. 3-13. Using your knowledge of the definitions, postulates, and theorems in this chapter, classify each of the following pairs of angles as one of the following: corresponding angles, alternate interior angles, same-side interior angles, alternate exterior angles, or same-side exterior angles.

(*a*) $\measuredangle 3$ and $\measuredangle 5$
(*b*) $\measuredangle 2$ and $\measuredangle 7$
(*c*) $\measuredangle 1$ and $\measuredangle 7$
(*d*) $\measuredangle 1$ and $\measuredangle 5$
(*e*) $\measuredangle 3$ and $\measuredangle 6$

(*f*) ∡4 and ∡6
(*g*) ∡4 and ∡8
(*h*) ∡1 and ∡8

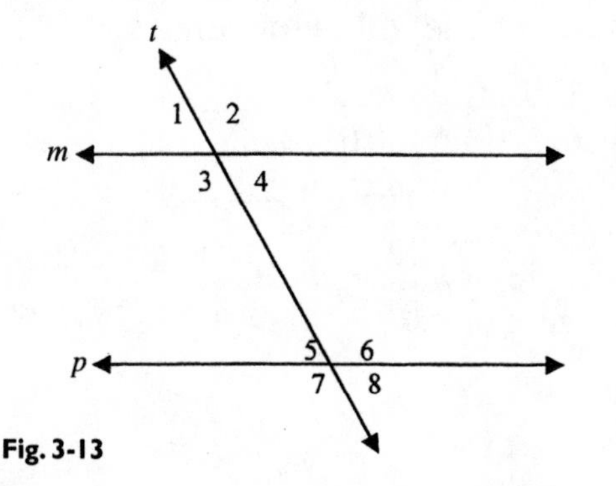

Fig. 3-13

Solution

(*a*) ∡3 and ∡5 are same-side interior angles because they are formed between the coplanar lines and they appear on the same side of the transversal.

(*b*) ∡2 and ∡7 are alternate exterior angles because they are formed outside of the coplanar lines and appear on opposite sides of the transversal.

(*c*) ∡1 and ∡7 are same-side exterior angles because they are formed outside of the coplanar lines and appear on the same side of the transversal.

(*d*) ∡1 and ∡5 are corresponding angles because they appear in the same position on the two lines cut by the transversal.

(*e*) ∡3 and ∡6 are alternate interior angles because they are formed inside of the coplanar lines and appear on opposite sides of the transversal.

(*f*) ∡4 and ∡6 are same-side interior angles because they are formed inside of the coplanar lines and appear on the same side of the transversal.

(*g*) ∡4 and ∡8 are corresponding angles because they appear in the same position on the lines cut by the transversal.

(*h*) ∡1 and ∡8 are alternate exterior angles because they are formed outside of the coplanar lines and appear on opposite sides of the transversal.

EXAMPLE 2

In a plane, two parallel lines r and s are cut by the transversal t, as you see in Fig. 3-14. Using your knowledge of the definitions, postulates, and theorems in this chapter, calculate the measures of the following angles.

(*a*) If $m\measuredangle 1 = 120°$, then $m\measuredangle 4 =$______.
(*b*) If $m\measuredangle 2 = 60°$, then $m\measuredangle 6 =$ ______.
(*c*) If $m\measuredangle 3 = 60°$, then $m\measuredangle 5 =$ ______.
(*d*) If $m\measuredangle 8 = 120°$, then $m\measuredangle 7 =$ ______.
(*e*) Find the value of $m\measuredangle 5 + m\measuredangle 6$.
(*f*) Find the value of $m\measuredangle 2 + m\measuredangle 7$, if $m\measuredangle 2 = 60°$.

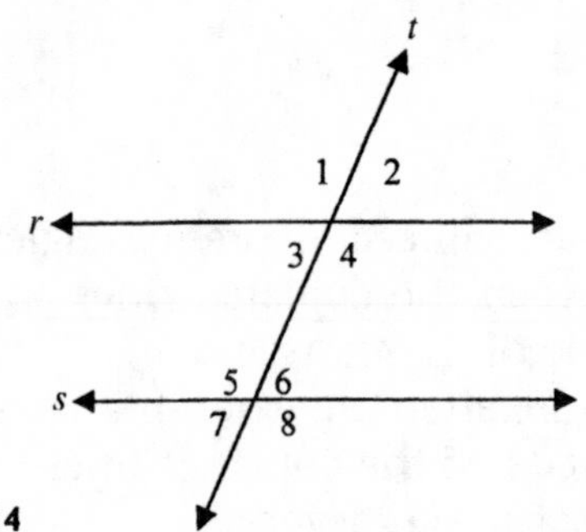

Fig. 3-14

Solution

(*a*) If $m\measuredangle 1 = 120°$, then $m\measuredangle 4 = 120°$, because they are vertical angles formed by the intersection of line r with the transversal t. Vertical angles are congruent and equal in measure.
(*b*) If $m\measuredangle 2 = 60°$, then $m\measuredangle 6 = 60°$ as well, because the two angles are corresponding angles, which are by definition congruent and equal in measure.
(*c*) If $m\measuredangle 3 = 60°$, then $m\measuredangle 5 = 120°$, because the two angles are same-side interior angles, which are supplementary. Thus, $m\measuredangle 3 + m\measuredangle 5 = 180°$. If $m\measuredangle 3 = 60°$, then its supplement $\measuredangle 5$ must measure 120°.
(*d*) If $m\measuredangle 8 = 120°$, then $m\measuredangle 7 = 60°$, because the two angles are supplementary. Thus, $m\measuredangle 8 + m\measuredangle 7 = 180°$. If $m\measuredangle 8 = 120°$, then its supplement $\measuredangle 7$ must measure 60°.

(*e*) The value of $m\measuredangle 5 + m\measuredangle 6$ is 180°, because the angles are supplementary.

(*f*) The value of $m\measuredangle 2 + m\measuredangle 7$ is 120°, because $\measuredangle 2$ and $\measuredangle 7$ are alternating exterior angles and have the same measure.

EXAMPLE 3

Find the value of x in each problem below, using the angle relationships as they appear in Fig. 3-15.

(*a*) If $m\measuredangle 2 = x + 20$ and $m\measuredangle 4 = 2x - 20$, what is the value of x?

(*b*) If $m\measuredangle 3 = x + 8$ and $m\measuredangle 5 = 3x + 12$, what is the value of x?

(*c*) If $m\measuredangle 6 = 3x - 15$ and $m\measuredangle 7 = 2x - 5$, what is the value of x?

(*d*) If $m\measuredangle 5 = 2x + 12$ and $m\measuredangle 8 = 4(x - 7)$, what is the value of x?

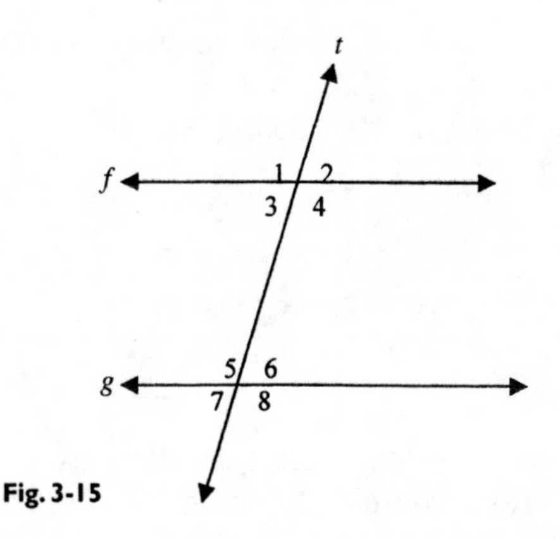

Fig. 3-15

Solution

(*a*) $\measuredangle 2$ and $\measuredangle 4$ are supplementary angles, so their sum is equal to 180°. To solve for x, the two angles must be added and their sum must be set equal to 180°, as below.

$$m\measuredangle 2 + m\measuredangle 4 = 180$$
$$(x + 20) + (2x - 20) = 180$$

$$x + 20 + 2x - 20 = 180$$
$$3x = 180$$
$$x = 60°$$

(*b*) $\measuredangle 3$ and $\measuredangle 5$ same-side interior angles, which are supplementary, so their sum is 180°. To solve for x, the two angles must be added and their sum set equal to 180° as below.

$$m\measuredangle 3 + m\measuredangle 5 = 180$$
$$(x + 8) + (3x + 12) = 180$$
$$x + 8 + 3x + 12 = 180$$
$$4x + 20 = 180$$
$$4x = 160$$
$$x = 40°$$

(*c*) $\measuredangle 6$ and $\measuredangle 7$ are vertical angles, so they are congruent and their measure is equal. To solve for x, set the two equations equal to each other as below.

$$m\measuredangle 6 = m\measuredangle 7$$
$$3x - 15 = 2x - 5$$
$$3x - 2x = 15 - 5$$
$$x = 10°$$

(*d*) $\measuredangle 5$ and $\measuredangle 8$ are vertical angles, so they are congruent and their measures are equal. To solve for x, set the equations equal to each other as below.

$$m\measuredangle 5 = m\measuredangle 8$$
$$2x + 12 = 4(x - 7)$$
$$2x + 12 = 4x - 28$$
$$12 + 28 = 4x - 2x$$
$$40 = 2x$$
$$10° = x$$

Supplementary Parallel Lines and Planes Problems

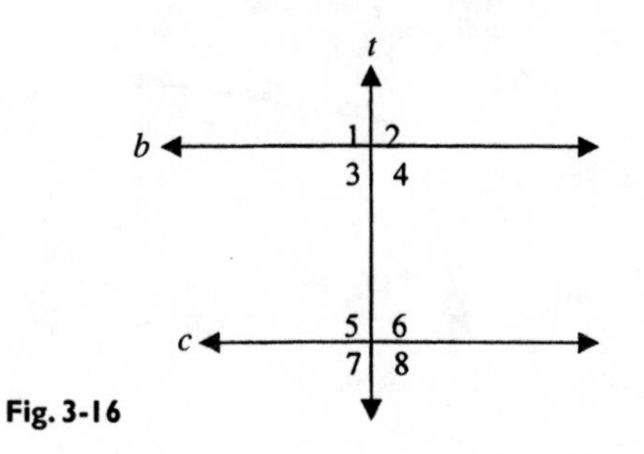

Fig. 3-16

Refer to Fig. 3-16 for Probs. 1 and 2.

1. If transversal t is perpendicular to line b, $m\measuredangle 4$ is ______.
2. If transversal t is perpendicular to line b and line b is parallel to line c, then $m\measuredangle 6$ is ______.

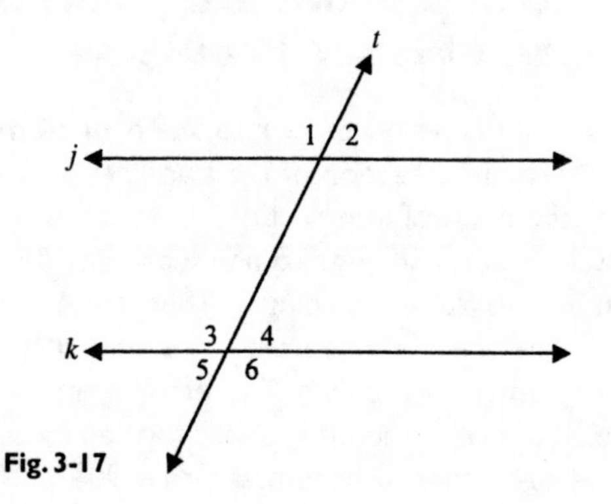

Fig. 3-17

Refer to Fig. 3-17 for Probs. 3 through 7.

3. If line j is parallel to line k and $m\measuredangle 2 = 75°$, then $m\measuredangle 4 =$ ______.
4. If line j is parallel to line k and $m\measuredangle 6 = 95°$, then $m\measuredangle 3 =$ ______.
5. If line j is parallel to line k and $m\measuredangle 2 = 75°$, then $m\measuredangle 1 =$ ______.
6. Find the value of $m\measuredangle 3 + m\measuredangle 5$.
7. Find the value of $m\measuredangle 2 + m\measuredangle 5$, if $m\measuredangle 2 = 75°$.

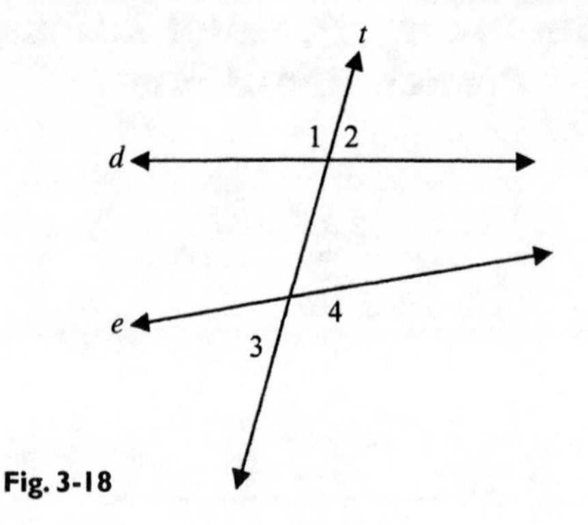

Fig. 3-18

Refer to Fig. 3-18 for Probs 8 through 10.

8. If $m\measuredangle 2 = 80°$, what is the value of $m\measuredangle 1$?
9. If $m\measuredangle 4$ is twice the value of $\measuredangle 3$, what are the measures of both $\measuredangle 3$ and $\measuredangle 4$?
10. Explain why the transversal *t* cannot be perpendicular to line *d* and line e.

Solutions to Supplementary Parallel Lines and Planes Problems

1. If the transversal *t* is perpendicular to line *b*, $m\measuredangle 4$ must be a right angle, or 90°. The intersection of the two lines forms perpendicular angles at the point of intersection.
2. If transversal *t* is perpendicular to line *b* and line *b* is parallel to line *c*, then $m\measuredangle 6$ is 90°. According to Theorem 4, if a transversal is perpendicular to one of two parallel lines, then it is also perpendicular to the other parallel line. Therefore, if transversal *t* is also perpendicular to line *c*, then the angles formed by their intersection must be right angles whose measure is 90°.
3. If line *j* is parallel to line *k* and $m\measuredangle 2 = 75°$, then $m\measuredangle 3 = 75°$ because $\measuredangle 2$ and $\measuredangle 4$ are corresponding angles and corresponding angles are congruent if formed when two parallel lines are cut by a transversal.
4. If $m\measuredangle 6 = 95°$, then $m\measuredangle 3$ is also 90° because $\measuredangle 6$ and $\measuredangle 3$ are vertical angles and vertical angles are equal in measure.
5. If $m\measuredangle 2 = 75°$, then $m\measuredangle 1 = 105°$ because $\measuredangle 2$ and $\measuredangle 1$ are supplementary angles and their sum is 180°.
6. The sum of $m\measuredangle 3$ and $m\measuredangle 5$ is 180° because $\measuredangle 3$ and $\measuredangle 5$ are sup-

plementary angles. The sum of supplementary angles must equal 180° by definition.

7. The sum of $m\measuredangle 2$ and $m\measuredangle 5$ is 150° because $\measuredangle 2$ and $\measuredangle 5$ are alternate exterior angles and have equal measure.
8. If $m\measuredangle 2 = 80°$, the value of $m\measuredangle 1$ is 100° because the angles are supplementary. Therefore, $m\measuredangle 2 + m\measuredangle 1 = 180°$. If $m\measuredangle 2 = 80°$, then $m\measuredangle 1 = 180 - 80 = 100°$.
9. If the value of $m\measuredangle 4$ is twice the value of $\measuredangle 3$, then we must use the following equation to calculate the measures of $\measuredangle 3$ and $\measuredangle 4$. From Fig. 3-18, we know that $\measuredangle 3$ and $\measuredangle 4$ are supplementary angles, so their sum is 180°.

$$2(m\measuredangle 3) = m\measuredangle 4$$
$$2(m\measuredangle 3) + m\measuredangle 4 = 180°$$

By substituting,

$$\tfrac{1}{2}(m\measuredangle 4) + m\measuredangle 4 = 180°$$
$$\tfrac{3}{2}\, m\measuredangle 4 = 180°$$
$$m\measuredangle 4 = \tfrac{2}{3}(180) = 120°$$

We were told that $2(m\measuredangle 3) = m\measuredangle 4$. Therefore,

$$2(m\measuredangle 3) = 120°$$
$$m\measuredangle 3 = 60°$$

10. The transversal t cannot be perpendicular to line d or line e because the angles formed by their intersection are not right (90°) angles.

Chapter 4

Congruent Triangles

You have already become familiar with congruent line segments (line segments that have equal length), and congruent angles (angles that have equal measures). In this chapter, you will combine the skills that you have developed and apply them to working with congruent triangles.

A triangle is a plane figure formed by joining three line segments at *noncollinear* points, meaning points not in a straight line. The point at which two segments join is called a *vertex* of the triangle, the plural being *vertices,* and the segments are the *sides* of the triangle.

Triangle ABC ($\triangle ABC$) appears in Fig. 4-1, with vertices at points A, B, and C. The sides of $\triangle ABC$ are identified as $\overline{AB}$, $\overline{BC}$, $\overline{CA}$. The angles of $\triangle ABC$ are identified as $\measuredangle A$, $\measuredangle B$, $\measuredangle C$.

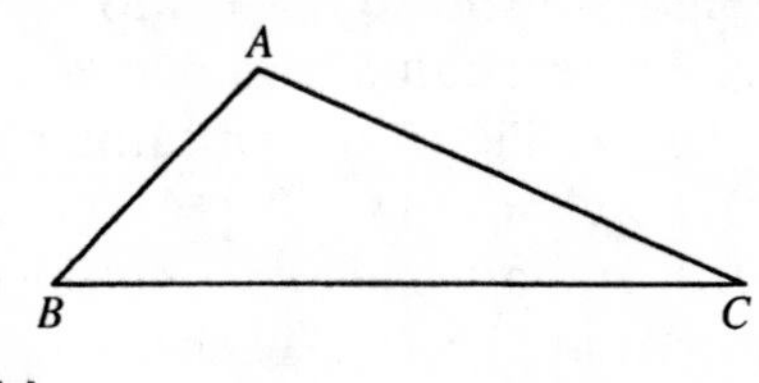

Fig. 4-1

Triangles are often classified according to the number of *congruent* sides they contain, as you may see in Figs. 4-2 through 4-4. By definition, a *scalene triangle* (Fig. 4-2) has no sides equal or congruent, an *isosceles triangle* (Fig. 4-3) contains two equal and congruent sides, and an *equilateral triangle* (Fig. 4-4) has all sides equal and congruent.

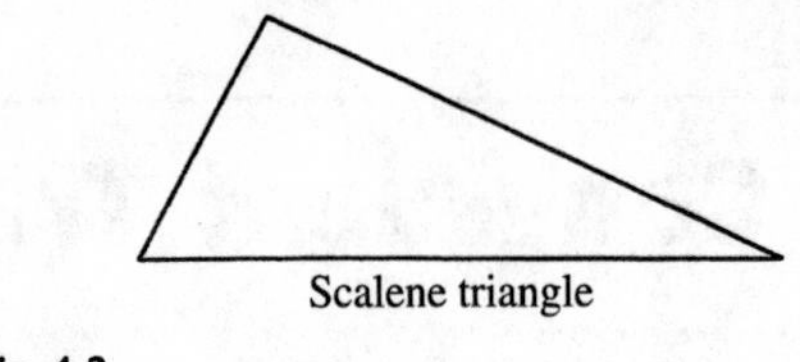

Fig. 4-2

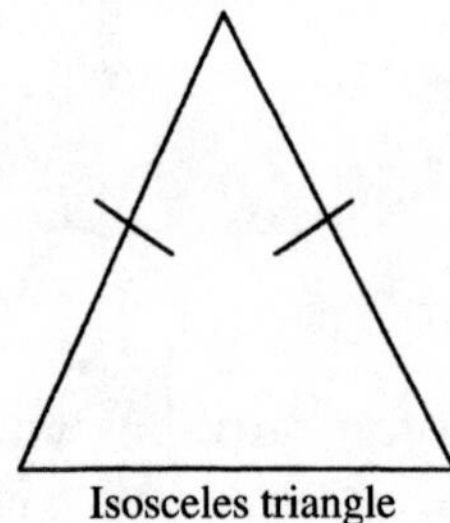

Fig. 4-3

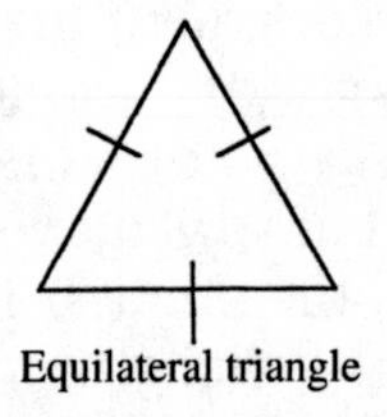

Fig. 4-4

Angles may also be used to name or classify triangles, as the examples in Figs. 4-5 through 4-8 show. By definition, an *acute triangle* (Fig. 4-5) contains three acute angles (less than 90°), an *obtuse triangle* (Fig. 4-6) contains one obtuse angle (greater than 90°), a *right triangle* (Fig. 4-7) contains one right angle (equal to 90°), and an *equiangular triangle* (Fig. 4-8) contains three congruent angles (60° each).

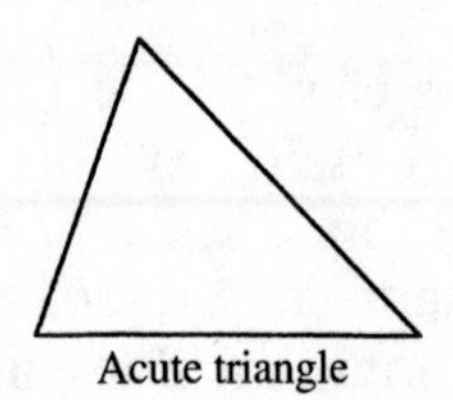

Fig. 4-5

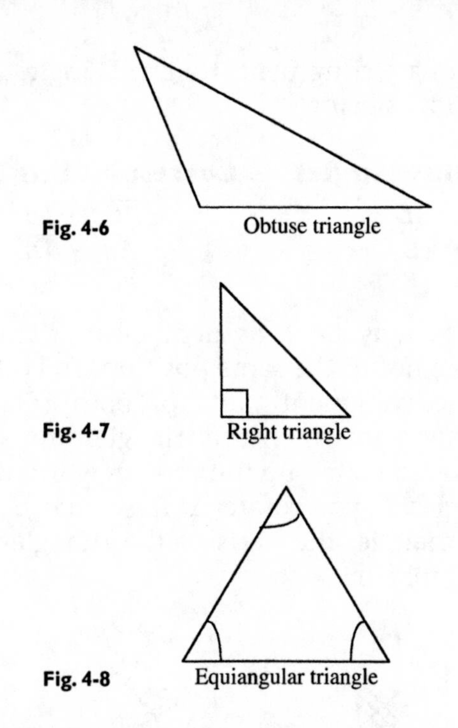

Fig. 4-6 Obtuse triangle

Fig. 4-7 Right triangle

Fig. 4-8 Equiangular triangle

Triangles are congruent if they have the same shape and if their corresponding angles and corresponding sides are equal. For one triangle to be congruent to another, their vertices must match so exactly in size that their corresponding parts fit perfectly on top of each other. To emphasize this congruence, we name the corresponding vertices in the same order, and we know that the corresponding sides of the triangles will also be equal. In short, corresponding parts of congruent triangles are congruent.

Triangles *ABC* and *DEF* in Fig. 4-9 are congruent, if we can place one on top of the other and line up the vertices so that they meet exactly. We indicate this congruence of the vertices in the following manner: $A \leftrightarrow D$, $B \leftrightarrow E$, $C \leftrightarrow F$.

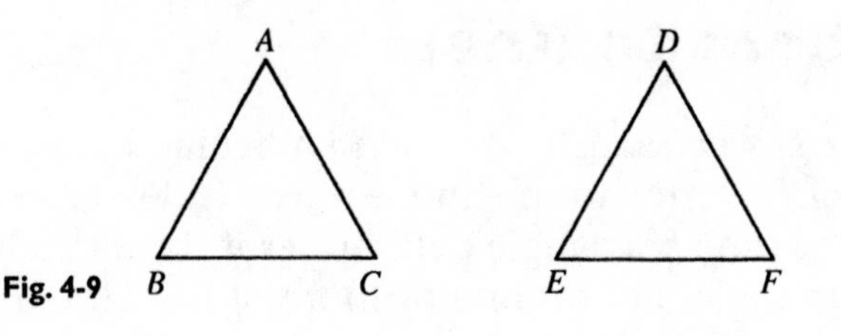

Fig. 4-9

If the vertices are congruent, then the angles and sides should correspond, as below:

Corresponding Angles	**Corresponding Sides**
$\measuredangle A \leftrightarrow \measuredangle D$	$\overline{AC} \leftrightarrow \overline{DF}$
$\measuredangle B \leftrightarrow \measuredangle E$	$\overline{AB} \leftrightarrow \overline{DE}$
$\measuredangle C \leftrightarrow \measuredangle F$	$\overline{BC} \leftrightarrow \overline{EF}$

Two triangles may be congruent even if their corresponding parts are not in the same positions in both figures. In Fig. 4-10, the congruent parts of both triangles are marked, but the way in which the triangles are positioned does not allow you to match up the vertices without rotating one of the triangles. If you rotate $\triangle KJL$ so that $\overline{JL}$ becomes the base of the triangle, the parts of the triangles that are marked alike will match.

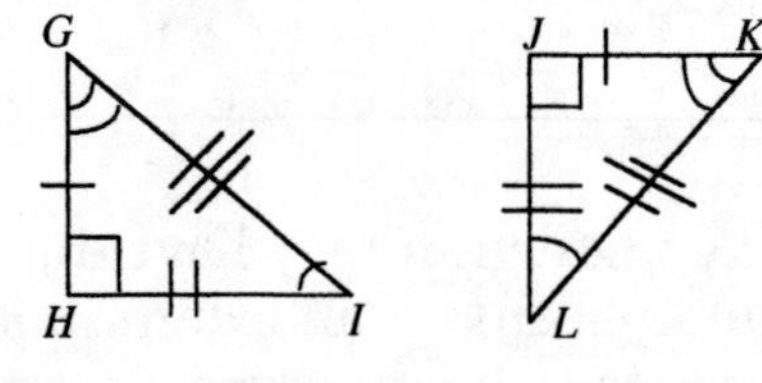

Fig. 4-10

The vertices of the triangles match as follows: $G \leftrightarrow K$, $H \leftrightarrow J$, $I \leftrightarrow L$. The corresponding sides are also congruent: $\overline{GH} \cong \overline{KJ}$, $\overline{HI} \cong \overline{JL}$, $\overline{GI} \cong \overline{KL}$. Because the corresponding parts of these triangles are congruent, the triangles are also congruent. To refer to congruent triangles, name the corresponding vertices in the same order. Thus, in Fig. 4-10, the following is true: $\triangle GHI$ is congruent to $\triangle KJL$ or $\triangle GHI \cong \triangle KJL$.

Definitions to Know

Altitude of a triangle. A perpendicular line segment drawn from a vertex to the line segment (side) opposite the vertex. The type of triangle determines if the altitude is inside the triangle, lies on one or more of the legs, or exists outside of the triangle. The altitudes of an acute triangle are shown in Fig. 4-11 and are always contained within the triangle.

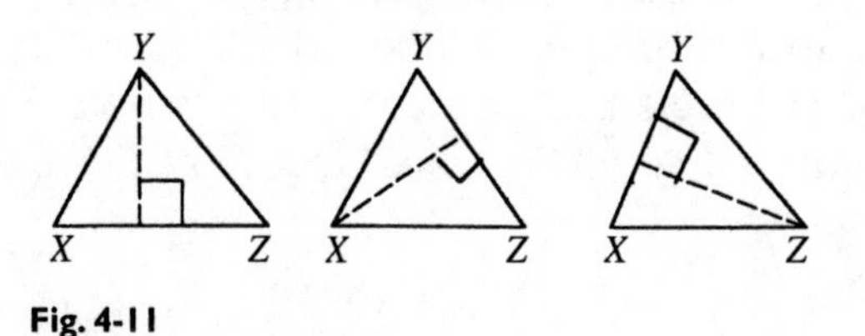

Fig. 4-11

Two of the altitudes of a right triangle are the legs, and the third altitude is contained within the triangle, as in Fig. 4-12.

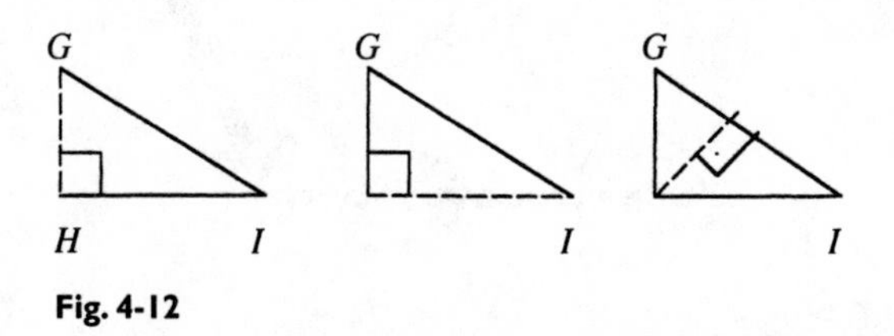

Fig. 4-12

Two of the altitudes of an obtuse triangle appear outside of the triangle, and the third altitude is inside the triangle, as in Fig. 4-13.

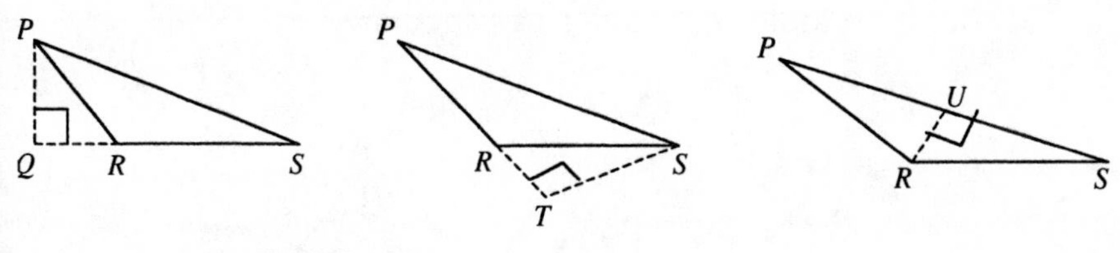

Fig. 4-13

Base of a triangle. The third side of an isosceles triangle, which contains two congruent sides.

Distance from a point to a line or plane. Defined as the length of the perpendicular segment drawn from that point to the line or plane.

Hypotenuse. The side opposite the right angle in a right triangle.

Legs of a triangle. The two congruent sides of an isosceles triangle or the two sides *not* opposite the right angle in a right triangle.

Line perpendicular to a plane. A line that intersects a plane and is perpendicular to all lines in the plane that pass through the point of intersection. In Fig. 4-14, $\overleftrightarrow{LO} \perp$ plane H, so $\overleftrightarrow{LO} \perp \overleftrightarrow{OA}$, $\overleftrightarrow{LO} \perp \overleftrightarrow{OB}$, $\overleftrightarrow{LO} \perp \overleftrightarrow{OC}$, and so forth.

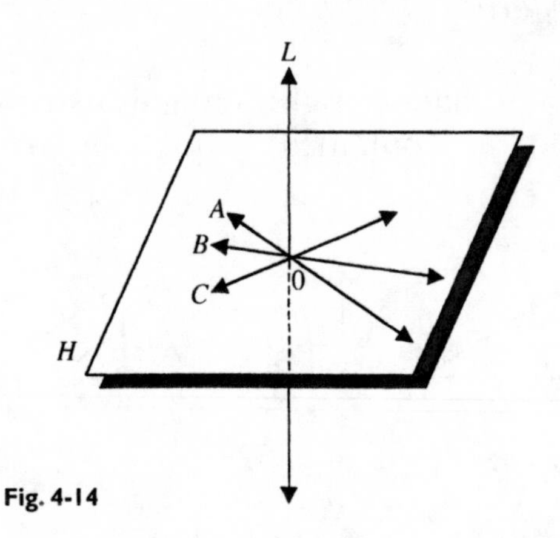

Fig. 4-14

Median. A line segment in a triangle that connects a vertex to the midpoint of the opposite side. As Fig. 4-15 shows, a triangle will have three medians.

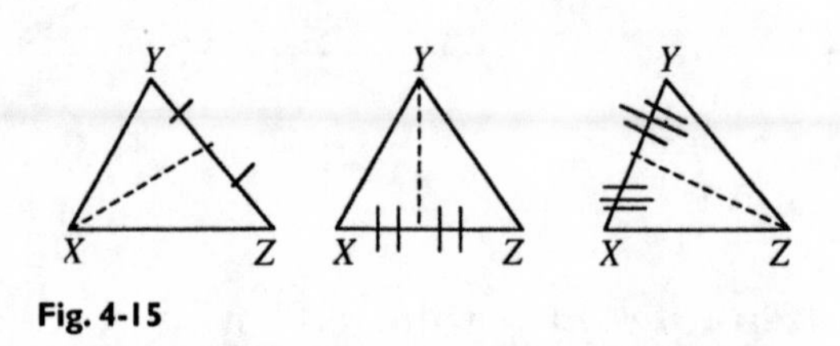

Fig. 4-15

Perpendicular bisector of a line segment. A line, ray, or line segment that is perpendicular to the segment at its midpoint. In Fig. 4-16, line *a* is the perpendicular bisector of $\overline{MP}$ and intersects the line segment at point M.

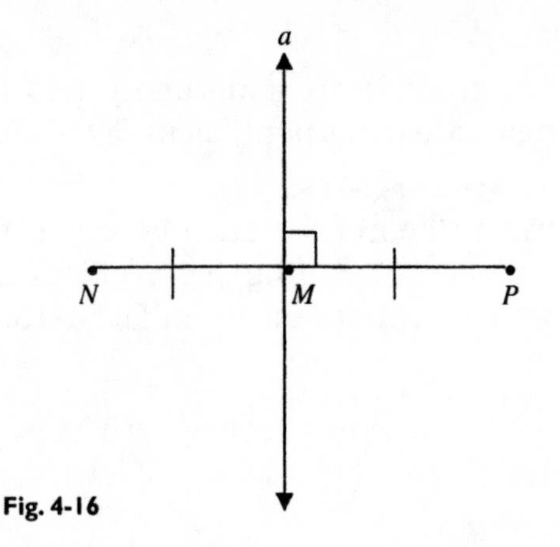

Fig. 4-16

Relevant Postulates and Theorems

Postulate 1 (Side-Side-Side)
If three sides of one triangle are congruent to three sides of another triangle, then the triangles are congruent.

Postulate 2 (Side-Angle-Side)
If two sides and the included angle of one triangle are congruent to two sides and the included angle of another triangle, then the triangles are congruent.

Postulate 3 (Angle-Side-Angle)
If two angles and the included side of one triangle are congruent to two angles and the included side of another triangle, then the triangles are congruent.

Theorem 1 (Isosceles Triangle)
If two sides of a triangle are congruent, then the angles opposite those sides are congruent, as in Fig. 4-17.

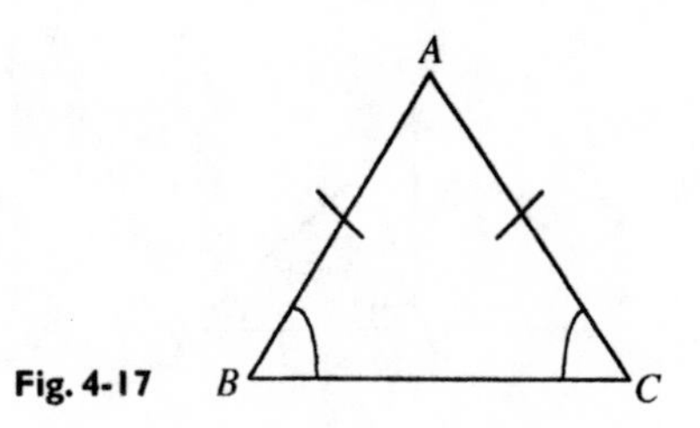

Fig. 4-17

Theorem 2

If two angles of a triangle are congruent, then the sides opposite those angles are congruent, as in Fig. 4-17.

Theorem 3 (Angle-Angle-Side)

If two angles and a side not included between the angles are congruent to the corresponding parts of another triangle, then the triangles are congruent, as in Fig. 4-18.

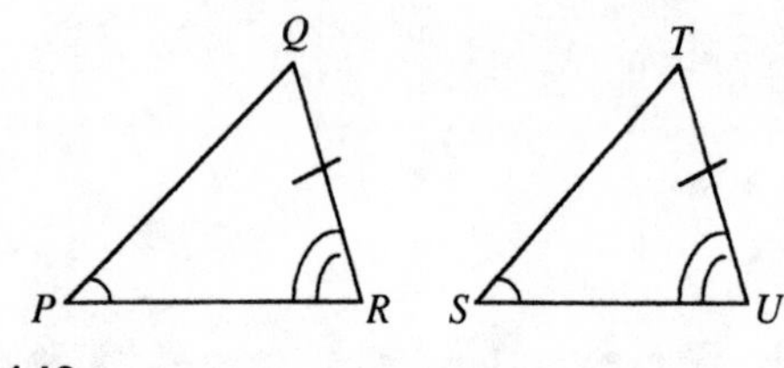

Fig. 4-18

Theorem 4 (Hypotenuse-Leg)

If the hypotenuse and a leg of one right triangle are congruent to the corresponding parts of another right triangle, then the triangles are congruent, as in Fig. 4-19.

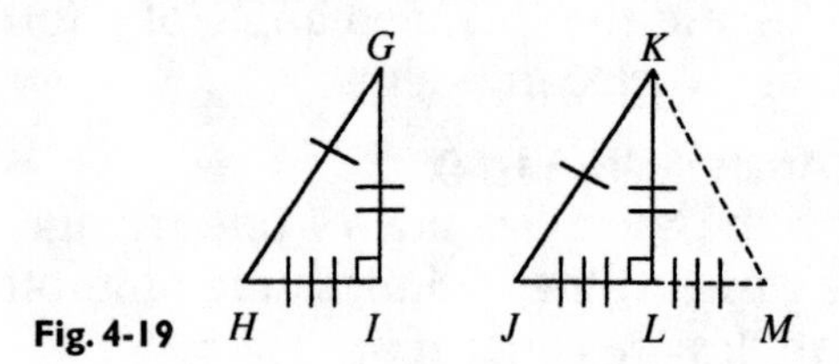

Fig. 4-19

Theorem 5

If a point lies on the perpendicular bisector of a segment, then the point is equidistant from the endpoints of the segment, as in Fig. 4-20.

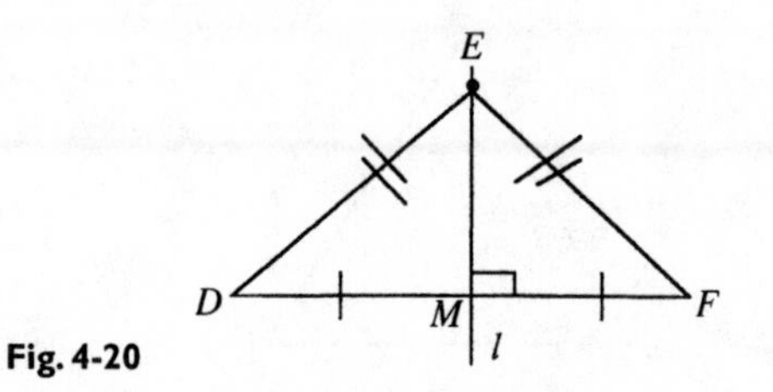

Fig. 4-20

Theorem 6

If a point is equidistant from the endpoints of a segment, then the point lies on the perpendicular bisector of the segment, as in Fig. 4-20.

Theorem 7

If a point lies on the bisector of an angle, then the point is equidistant from the sides of the angle, as in Fig. 4-21.

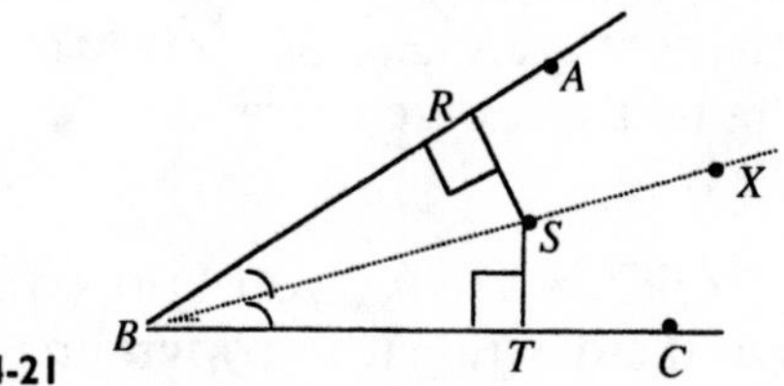

Fig. 4-21

Theorem 8

If a point is equidistant from the sides of an angle, then the point lies on the bisector of the angle, as in Fig. 4-21.

EXAMPLE 1

Using the definition of congruent triangles, list six congruences that exist if $\triangle FGH \cong \triangle IJK$.

Solution

According to the definition of congruent triangles, corresponding parts of congruent triangles are congruent. Also remember that we refer to congruent triangles by naming the corresponding vertices in identical order. Therefore, the congruent triangles in this example should appear as in Fig. 4-22.

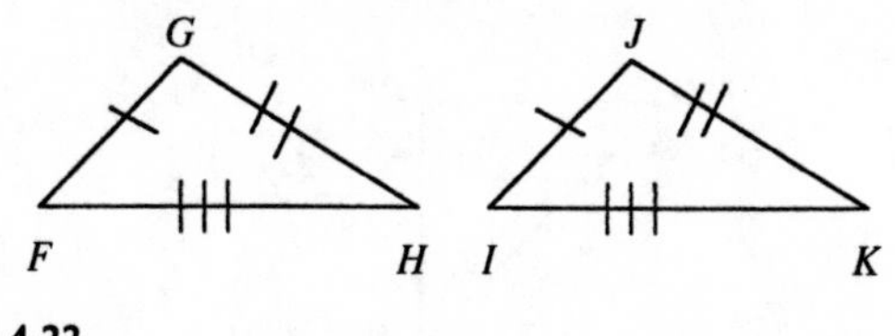

Fig. 4-22

Both the angles and the sides of one triangle must correspond to the angles and the sides of the second triangle. Thus, the following congruencies can be justified:

Congruent Angles	**Congruent Sides**
$\measuredangle F \leftrightarrow \measuredangle I$	$\overline{FG} \cong \overline{IJ}$
$\measuredangle G \leftrightarrow \measuredangle J$	$\overline{GH} \cong \overline{JK}$
$\measuredangle H \leftrightarrow \measuredangle K$	$\overline{FH} \cong \overline{IK}$

EXAMPLE 2

If $\triangle ABC$ is congruent to $\triangle DEF$, and the measure of $\measuredangle \mathrm{A} = 80°$ while the measure of $\measuredangle F = 60°$, identify the measures of the following angles: $\measuredangle B$, $\measuredangle C$, $\measuredangle D$, $\measuredangle E$. How can you be certain that the measures are correct?

Solution

Given that $\triangle ABC \cong \triangle DEF$, you know that the corresponding angles and sides must be congruent, according to the definition of congruent triangles. As a result, $\measuredangle A \cong \measuredangle D$, $\measuredangle B \cong \measuredangle E$, and $\measuredangle C \cong \measuredangle F$. We already know that $m\measuredangle A = 80°$ and $m\measuredangle F = 60°$, so the measures of their corresponding angles, $\measuredangle D$ and $\measuredangle C$, respectively, are the same.

Using the definition of a triangle, we can calculate the measure of the third angle in each of the two congruent triangles:

$$m\measuredangle A + m\measuredangle B + m\measuredangle C = 180°$$

$$m\measuredangle D + m\measuredangle E + m\measuredangle F = 180°$$

Therefore, the following must be true: $m\measuredangle D = 80°$, $m\measuredangle C = 60°$, $m\measuredangle B = 40°$, and $m\measuredangle E = 40°$.

EXAMPLE 3

Write a two-column proof for Fig. 4-23.

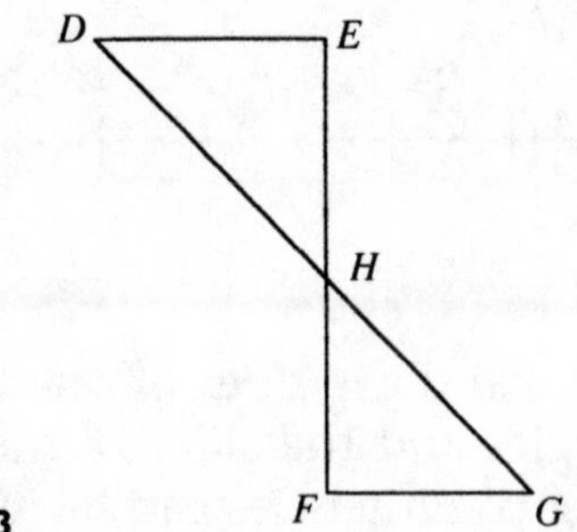

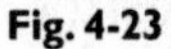
Fig. 4-23

Given:

$$\overline{DE} \perp \overline{EF}; \overline{FG} \perp \overline{EF}; H \text{ is the midpoint of } \overline{EF}$$

Prove:

$$\triangle DEH \cong \triangle GFH$$

Solution

Statements	Reasons
1. $\overline{DE} \perp \overline{EF}; \overline{FG} \perp \overline{EF}$	1. Given
2. $m\measuredangle E = 90°; m\measuredangle F = 90°$	2. Definition of perpendicular lines
3. $\measuredangle E \cong \measuredangle F$	3. Definition of congruent angles
4. H is the midpoint of $\overline{EF}$	4. Given
5. $\overline{EH} = \overline{HF}$	5. Definition of midpoint
6. $\measuredangle DHE \cong \measuredangle GHF$	6. Vertical angles are congruent.
7. $\triangle DEH \cong \triangle GFH$	7. Angle-Side-Angle Postulate

EXAMPLE 4

Provide a reason for each step in the following proof for Fig. 4-24.

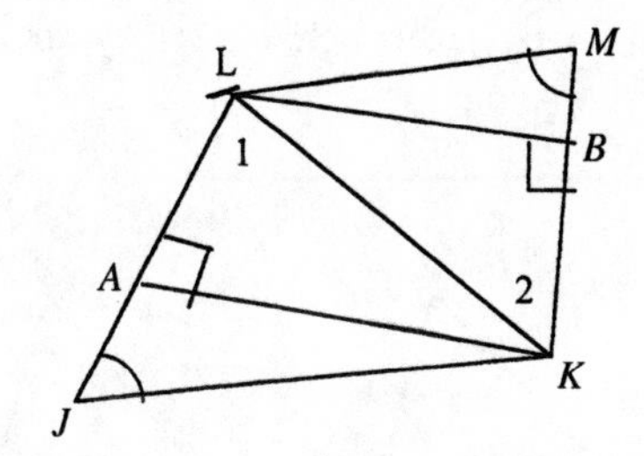

Fig. 4-24

Given:

$$\overline{JX} \cong \overline{MY}; \measuredangle J \cong \measuredangle M; \overline{KX} \perp \overline{JL}; \overline{LB} \perp \overline{KM}$$

Prove:

$$\overline{LJ} \parallel \overline{MK}$$

Statements	Reasons
1. $\triangle JKA \cong MLB$	1.
2. $\overline{KA} \cong \overline{LB}$	2.
3. $\triangle KLA \cong \triangle LKB$	3.
4. $\measuredangle 1 \cong \measuredangle 2$	4.
5. $\overline{LJ} \parallel \overline{MK}$	5.

Solution

Statements	Reasons
1. $\triangle JKA \cong MLB$	1. Postulate 3 (Angle-Side-Angle)
2. $\overline{KA} \cong \overline{LB}$	2. Corresponding parts of congruent triangles
3. $\triangle KLA \cong \triangle LKB$	3. Theorem 4 (Hypotenuse-Leg)
4. $\measuredangle 1 \cong \measuredangle 2$	4. Corresponding parts of congruent triangles
5. $\overline{LJ} \parallel \overline{MK}$	5. Two lines are parallel if they are cut by a transversal and their alternate interior angles are congruent.

Supplementary Congruent Triangles Problems

Given that $\triangle QRS$ is congruent to $\triangle XYZ$ (Fig. 4-25), complete the following statements in Probs. 1 through 5. Justify your answer with a definition, postulate, or theorem.

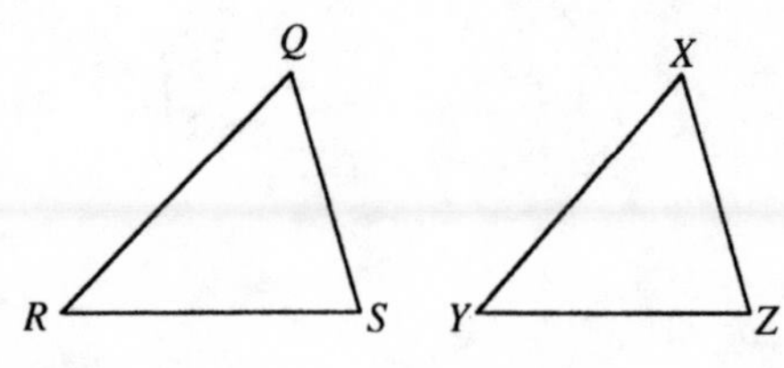

Fig. 4-25

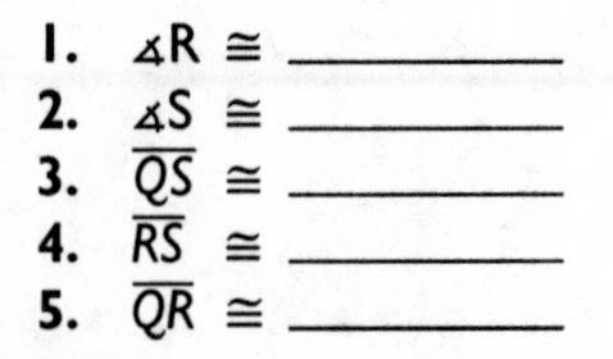

1. $\measuredangle R \cong$ ________

2. $\measuredangle S \cong$ ________

3. $\overline{QS} \cong$ ________

4. $\overline{RS} \cong$ ________

5. $\overline{QR} \cong$ ________

Given that the triangles in Fig. 4-26 are congruent, answer Probs. 6 through 10.

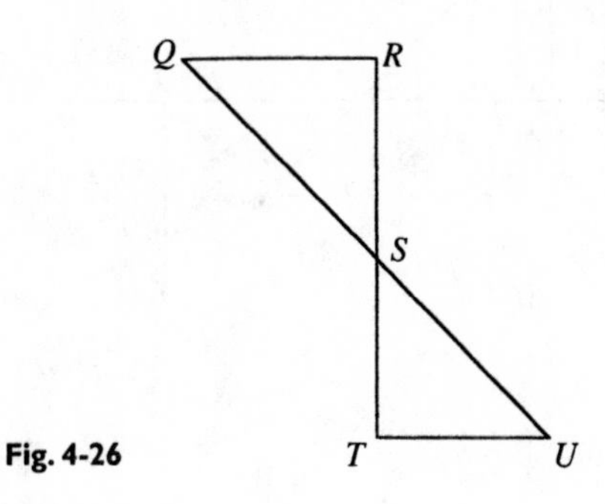

Fig. 4-26

6. $\triangle QRS \cong$ __________
7. $\overline{TS} \cong$ __________
8. $\measuredangle U \cong$ __________
9. Explain how you can determine that S is the midpoint of any segment.
10. $QU =$ __________
11. Write a two-column proof for Fig. 4-27.

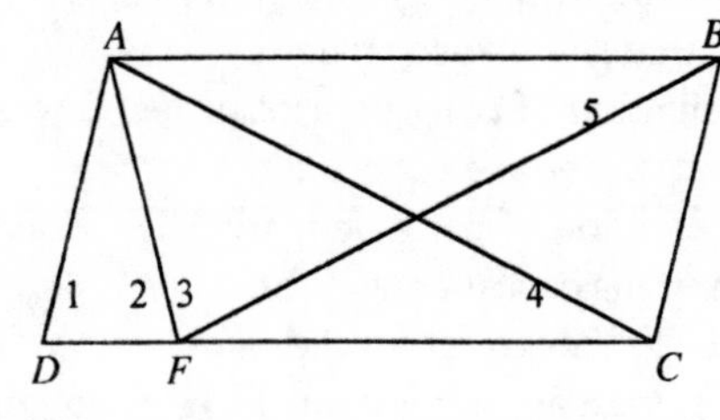

Fig. 4-27

Given:

$$\measuredangle 1 \cong 2 \cong 3; \overline{FB} \cong \overline{DC}$$

Prove:

$$\measuredangle 4 \cong \measuredangle 5$$

12. Write a two-column proof for Fig. 4-28, given the congruences as marked.

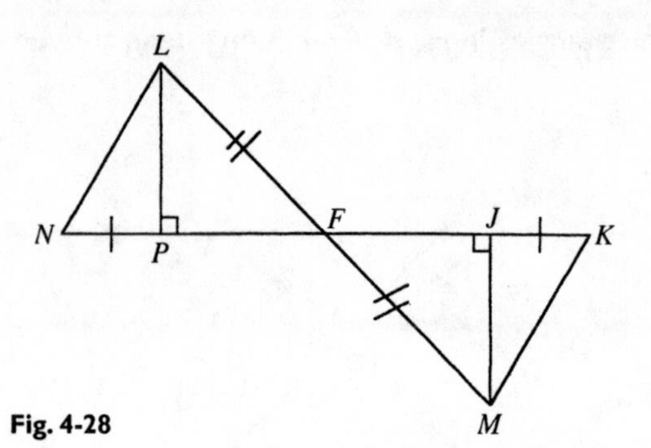

Fig. 4-28

Prove:

$$\measuredangle N \cong \measuredangle K$$

Solutions to Congruent Triangles Problems

1. $\measuredangle R \cong \measuredangle Y$. Definition of congruent triangles: Corresponding angles of congruent triangles are congruent.
2. $\measuredangle S \cong \measuredangle Z$. Definition of congruent triangles: Corresponding angles of congruent triangles are congruent.
3. $\overline{QS} \cong \overline{XZ}$. Definition of congruent triangles: Corresponding sides of congruent triangles are congruent.
4. $\overline{RS} \cong \overline{YZ}$. Definition of congruent triangles: Corresponding sides of congruent triangles are congruent.
5. $\overline{QR} \cong \overline{XY}$. Definition of congruent triangles: Corresponding sides of congruent triangles are congruent.
6. $\triangle QRS \cong \triangle UTS$. You are given the information that the triangles in Fig. 4-26 are congruent. To refer to congruent triangles, you must name the corresponding vertices in the same order.
7. $\overline{TS} \cong \overline{RS}$. Definition of congruent triangles: Corresponding sides of congruent triangles are congruent.
8. $\measuredangle U \cong \measuredangle Q$. Definition of congruent triangles: Corresponding angles of congruent triangles are congruent.
9. S is the midpoint of both $\overline{QU}$ and $\overline{RT}$ based on the definition of midpoint. The problem states that the two triangles are congruent, and according to the definition of congruent triangles, the corresponding parts of the triangles are congruent. Because of this, $\overline{QS}$ is congruent to $\overline{SU}$. Therefore, S is the midpoint of $\overline{QU}$. Based on

the definition of congruent triangles, $\overline{RS}$ is congruent to $\overline{ST}$. Therefore, S is also the midpoint of $\overline{RT}$.

10. $\overline{QU} = \overline{RT}$. Corresponding parts of congruent triangles are equal.

11.

Statements	Reasons
1. $\measuredangle 1 \cong \measuredangle 2 \cong \measuredangle 3$	1. Given
2. $\overline{AD} \cong \overline{AF}$	2. Theorem 1 (Isosceles Triangle Theorem)
3. $\overline{DC} \cong \overline{FB}$	3. Given
4. $\triangle ADC \cong \triangle AFB$	4. Postulate 2 (Side-Angle-Side)
5. $\measuredangle 4 \cong \measuredangle 5$	5. Definition of congruent triangles

12.

Statements	Reasons
1. $\triangle LFK \cong \triangle MFJ$	1. Theorem 3 (Angle-Angle-Side)
2. $\overline{LP} \cong \overline{MJ}$	2. Corresponding sides of congruent triangles
3. $\triangle NLP \cong \triangle KMJ$	3. Postulate 2 (Side-Angle-Side)
4. $\triangle N \cong \triangle K$	4. Corresponding parts of congruent triangles

Chapter 5

Quadrilaterals

A simple definition of a *quadrilateral* is that it is a polygon with four sides in which the sum of the interior angles is 360°. You might observe quadrilaterals that have all sides of different lengths, two sides of the same length and two of different lengths, or all four sides of the same length. All that the definition requires are the following: (1) Each segment must intersect exactly with two other segments, one at each endpoint, and (2) no two segments that have a common endpoint may be collinear.

In studying geometry, we explore relationships, so we study figures that allow us to observe meaningful similarities or differences. For this reason, a quadrilateral that has four sides of different length is of little value to our study unless we can subdivide it into familiar figures to which we can apply the principles of measurement that we have learned. In contrast, quadrilaterals such as the *parallelogram,* which has both pairs of opposite sides parallel, or the *trapezoid,* which has only one set of sides parallel, hold a great deal of interest. The definitions, postulates, and theorems related to parallel lines and triangles are especially useful in solving problems involving parallelograms and trapezoids.

Definitions to Know

Bases of a trapezoid. The two parallel sides of a trapezoid, as identified in Fig. 5-1.

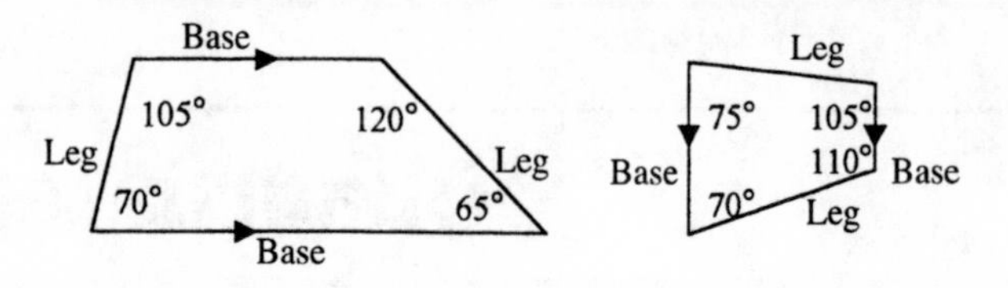

Fig. 5-1

Diagonal. A line segment that joins two nonconsecutive vertices of a polygon.

Isosceles trapezoid. A trapezoid with congruent legs and congruent base angles, as exhibited in Fig. 5-2.

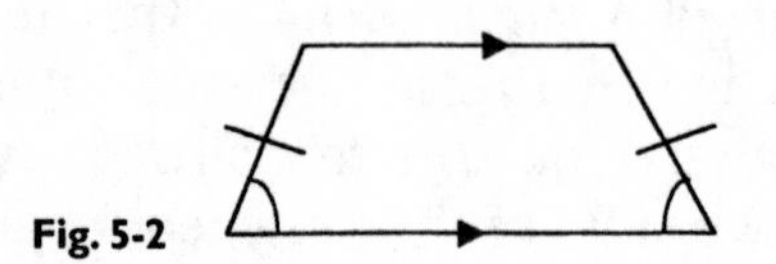
Fig. 5-2

Legs of a trapezoid. The two nonparallel sides of a trapezoid, as identified in Fig. 5-1.

Median of a trapezoid. A line segment parallel to the two bases that connects the midpoints of the two trapezoid legs and that is equal in length to the average of the two base lengths, as identified in Fig. 5-3.

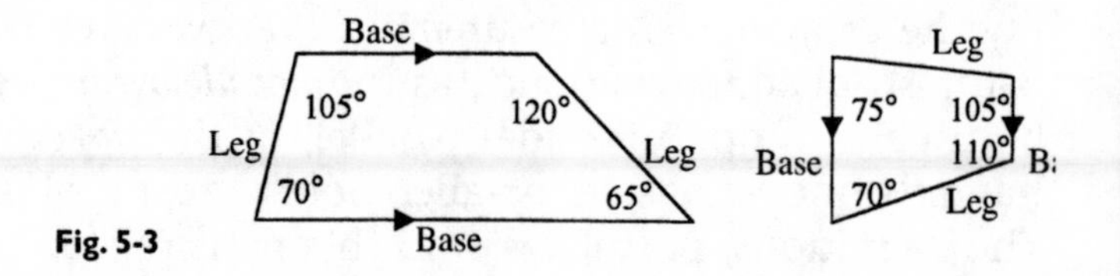

Fig. 5-3

Rectangle. A quadrilateral that is a parallelogram with four right angles and diagonals that are congruent.

Rhombus. A quadrilateral that is a parallelogram with four congruent sides and diagonals that are perpendicular.

Square. A quadrilateral that is a parallelogram with four right angles and four congruent sides.

Relevant Theorems

Theorem 1

Opposite sides and opposite angles of a parallelogram are congruent.

Theorem 2

The diagonals of a parallelogram bisect each other. (See Fig. 5-4.)

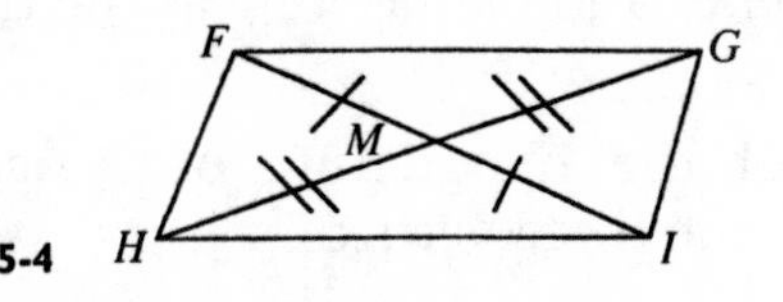

Fig. 5-4

Theorem 3

If both pairs of opposite sides or opposite angles of a quadrilateral are congruent, then the quadrilateral is a parallelogram.

Theorem 4

If one pair of opposite sides of a quadrilateral are both congruent and parallel, then the quadrilateral is a parallelogram.

Theorem 5

If two lines are parallel, then all points on one line are equidistant from all corresponding points on the other line. (See Fig. 5-5.)

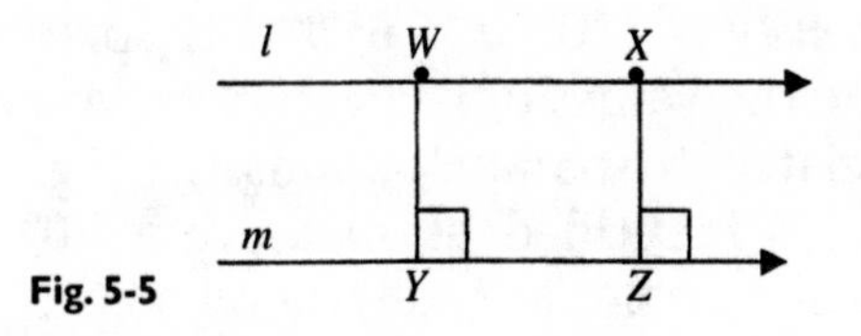

Fig. 5-5

Theorem 6

A line that contains the midpoint of one side of a triangle and is parallel to another side passes through the midpoint of the third side. (See Fig. 5-6.)

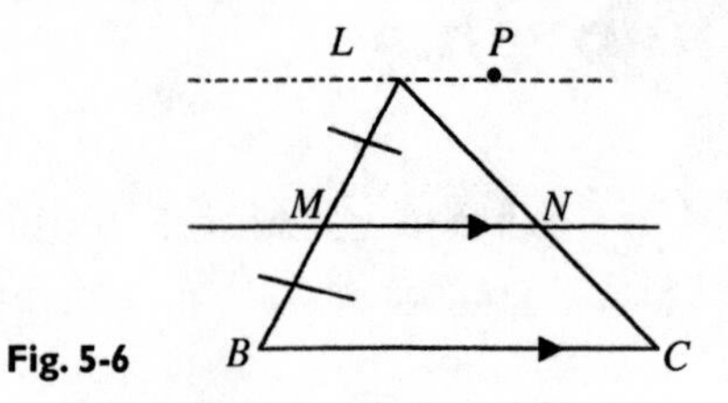

Fig. 5-6

Theorem 7

The line segment that joins the midpoints of two sides of a triangle is parallel to and half as long as the third side.

Theorem 8

The midpoint of the hypotenuse of a right triangle is equidistant from the three vertices.

EXAMPLE 1

If quadrilateral *ABCD* is a parallelogram, find the values of the variables in Fig. 5-7 and explain your answers. What is the perimeter of the figure?

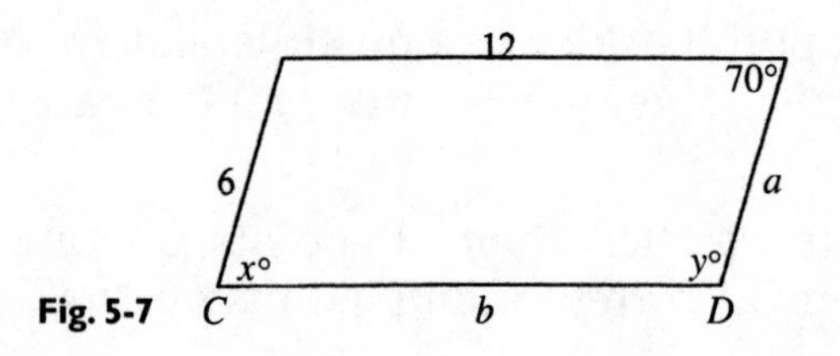

Fig. 5-7

Solution

$m\measuredangle x = 70°$; $m\measuredangle y = 110°$; $a = 6$; $b = 12$; perimeter = 36.

To determine the values of the variables, you must apply the theorems related to parallelograms. In Fig. 5-7, $\measuredangle x$ is opposite the angle identified as measuring 70°. Based on Theorem 1, opposite angles are congruent, so the measure of $\measuredangle x = 70°$. Because the figure is a parallelogram, the remaining two opposite angles are congruent as well. Based on the definition of quadrilaterals, the sum of the interior angles is 360°. You have already identified the sum of one set of opposite angles as being $70° + 70° = 140°$. Thus, the sum of the remaining angles can only be $360° - 140° = 220°$. If the measure of $\measuredangle y$ is one-half that amount, then based on Theorem 1 the measure of $\measuredangle y = 110°$.

To determine the values of the variables that represent the sides of the parallelogram, remember that the opposite sides in a parallelogram are also congruent. Therefore, $a = 6$, because the two sides are opposite each other and $b = 12$ for the same reason.

The perimeter of $\square ABCD$ is the sum of the four sides. Therefore, the perimeter of $\square ABCD = 12 + 6 + 12 + 6 = 36$.

EXAMPLE 2

If *BEAR* is a parallelogram with dimensions as shown in Fig. 5-8, find the lengths of the diagonals and the measures of the sides indicated by variables. Explain your answers.

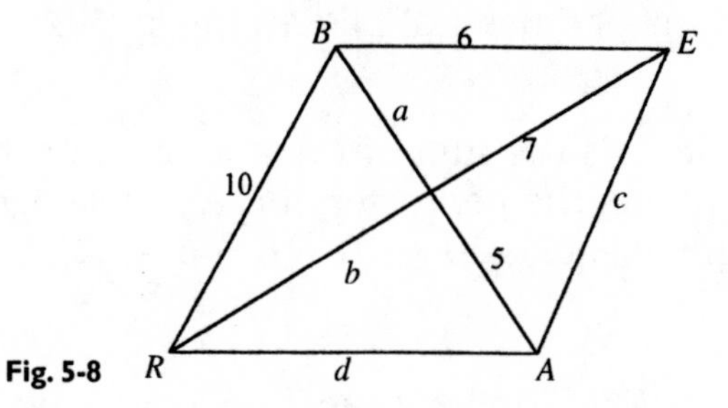

Fig. 5-8

Solution

The two diagonals measure 14 and 10; $c = 10$; $d = 6$.

The problem states that *BEAR* is a parallelogram. By definition, a diagonal is a line segment that joins two non-consecutive vertices of a polygon. Therefore, $\overline{BA}$ and $\overline{ER}$ are diagonals of the parallelogram *BEAR*. According to Theorem 2 the diagonals of a parallelogram bisect each other, so the length of segment b is equal in measure to 7. Concurrently, the length of segment a is equal in measure to 5.

The question does not, however, ask for the values of the variables. Instead, you are asked to identify the lengths of the diagonals and the measures of the sides that are labeled with variables. Therefore, you must add the values for the two halves of each diagonal to compute the total measure of each diagonal. Therefore, the diagonal formed by $\overline{BA} = b + 7$, but you determined that $b = 7$. So, the diagonal $\overline{BA} = 7 + 7 = 14$. Compute the measure of line segment *ER*, which forms the second diagonal in ▱*BEAR*. You have already determined that *a* is equal to 5, but the diagonal is the sum of the two numbers. You must also add the values for the two halves of the diagonal formed by $\overline{ER}$. Therefore, the second diagonal is equal to $a + 5 = 5 + 5 = 10$.

Side *c* is opposite the side having a measure of 10, so *c* is also equal to 10 based on Theorem 1, which states that the opposite sides of a parallelogram are congruent. Side *d* is opposite the side having a measure of 6, so *d* is also equal to 10 based on Theorem 1.

EXAMPLE 3

Calculate the perimeter of ▱*BEAR* in Fig. 5-8.

Solution

The perimeter is the sum of the four sides of the parallelogram. Therefore, the perimeter of the parallelogram that appears in Fig. 5-8 equals $10 + 10 + 6 + 6 = 32$.

EXAMPLE 4

In rhombus *KELP* that appears in Fig. 5-9, the measure of $\measuredangle 1$ equals 35°. Find the measures of the following angles: $\measuredangle 2$, $\measuredangle 3$, $\measuredangle 4$, $\measuredangle 5$.

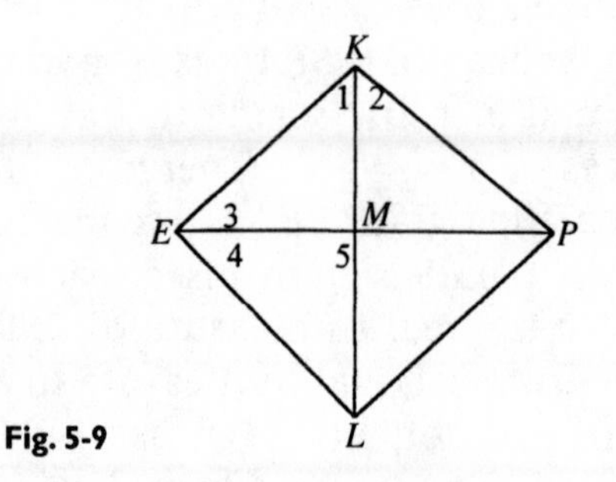

Fig. 5-9

Solution

The measures of the angles are as follow: $\measuredangle 2 = 35°$; $\measuredangle 3 = 55°$; $\measuredangle 4 = 55°$; $\measuredangle 5 = 90°$.

According to the definition of a rhombus, the diagonals are perpendicular to each other, so they form four right angles at the point they intersect. Therefore, $\measuredangle 5$ and each of the other three angles formed at the intersection of the two diagonals equal 90°. If $\measuredangle 1 = 35°$, then $\measuredangle 3$ must equal 55°, because $\measuredangle 1$ and $\measuredangle 3$ are the two acute angles in the right triangle *KME*.

To determine the measures of $\measuredangle 2$, use what you know about corresponding parts of congruent triangles. First, you know that $\overline{KE} \cong \overline{KP}$ because the sides of a rhombus are of the same measure. You also know that $\overline{EM} \cong \overline{MP}$, because the diagonals of a parallelogram bisect each other, and $\overline{KM}$ is equal to itself. Further, $\measuredangle KME = \measuredangle KMP = 90°$ because these angles are formed by the intersection of the diagonals, which in a rhombus results in right angles. So you can safely conclude that the two triangles, $\triangle KME$ and $\triangle KMP$, are congruent. Therefore, their corresponding parts are also congruent. As a result, then $\measuredangle 1 = \measuredangle 2 = 35°$.

Approach the calculation of the measure of $\measuredangle 4$ in the same manner. Based on the definition of a rhombus, $\overline{LE} \cong \overline{KE}$ because they are the sides of a rhombus. You also know that $\overline{KM} \cong \overline{LM}$ because the diagonals of a rhombus bisect each other, and $\overline{ME}$ is equal to itself. Finally, $\measuredangle KME = \measuredangle LME = 90°$ because these angles are formed by the intersection of the diagonals, which in a rhombus results in right angles. Thus, because three sides are congruent, we can safely conclude that the $\triangle LME \cong \triangle KME$. And, because corresponding parts of congruent triangles are congruent, $\measuredangle 3 = \measuredangle 4 = 55°$.

EXAMPLE 5

The isosceles trapezoid in Fig. 5-10 represents a plot of land which the owner wants to divide at the median as in-

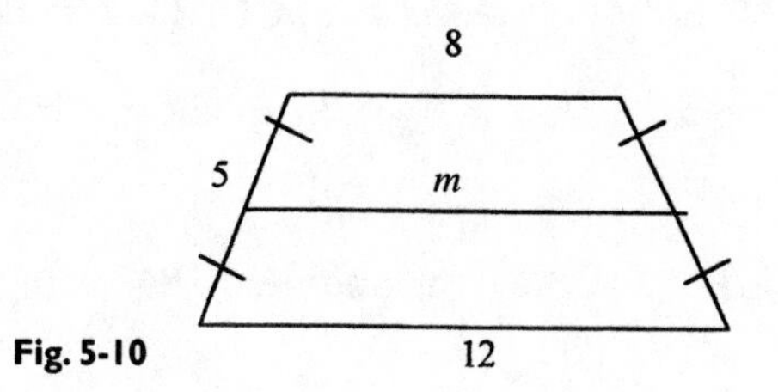

Fig. 5-10

dicated and enclose within a fence that will surround the entire property. Determine both the length of the divider and the amount of fencing materials that the owner must purchase.

Solution

The length of the divider in Fig. 5-10 is the measure of the median. Based on definition, the measure of the median of a trapezoid is the average of the sum of the two bases of the trapezoid. Therefore, the length of the divider should be calculated by adding the bases and dividing by 2, as follows: $(8 + 12)/2 = 10$.

To determine the amount of fencing material that the owner will need, you must calculate the perimeter of the trapezoid. According to definition, the legs of an isosceles trapezoid are equal in length. We must, therefore, add the measures of the congruent legs to the measures of the bases to find the amount of fencing required, as follows: $10 + 10 + 8 + 12 = 40$.

EXAMPLE 6

If the measure of one base angle of the isosceles trapezoid in Fig. 5-10 is equal to 70°, what are the measures of the remaining three angles?

Solution

The base angles in an isosceles trapezoid are congruent, according to definition. Therefore, if one base angle is equal to 70°, then so is the other base angle. Further, because the sum of the interior angles of a quadrilateral equals 360°, the sum of the two remaining angles of the trapezoid in Fig. 5-10 is $360° - 140° = 220°$. Therefore, the measures of the angles of the trapezoid are 70°, 70°, 110°, and 110°.

Supplementary Quadrilaterals Problems

Refer to the right trapezoid *TRAP* in Fig. 5-11 to answer Probs. 1 through 5.

1. What is the value of x in the right trapezoid *TRAP* with a median of $\overline{MD}$ in Fig. 5-11?

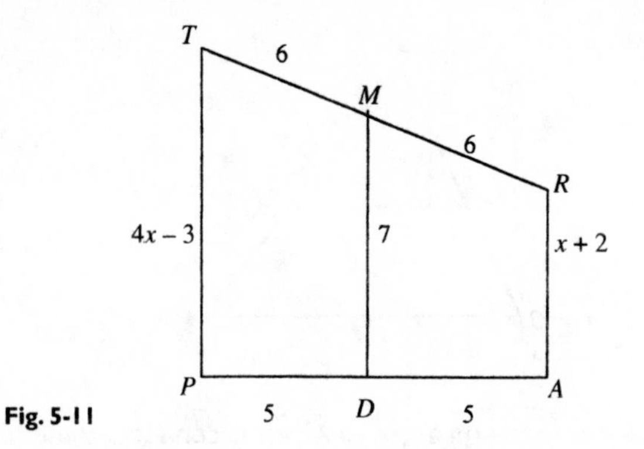

Fig. 5-11

2. What are the measures of the bases of right trapezoid *TRAP?*
3. What is the perimeter of right trapezoid *TRAP?*
4. How much will the measure of the median increase if x is equal to 7?
5. If the measure of $\measuredangle TPA = \measuredangle PAR = 90°$ and the measure of $\measuredangle PTR = 80°$, what is the measure of $\measuredangle TRA$?

Refer to the parallelogram *SEAL* in Fig. 5-12 to answer Probs. 6 through 10.

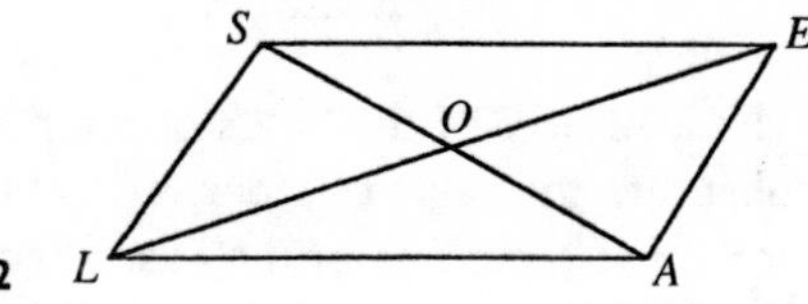

Fig. 5-12

6. If the measure of $\overline{EO}$ is equal to 7, what is the measure of diagonal $\overline{EL}$?
7. If the measure of diagonal $\overline{SA}$ is 10, what is the value of $\overline{SO}$?
8. If the measure of $\measuredangle SEL = 30°$, what is the measure of $\measuredangle ELA$?
9. What is the measure of $\measuredangle EOA$ if the measure of $\measuredangle SOL = 75°$?
10. If the perimeter of $\square SEAL$ is equal to 34 and the length of side $\overline{EA}$ is equal to 6, what is the measure of $\overline{AL}$?

Refer to $\triangle CAT$ in Fig. 5-13 to answer Probs. 11 and 12.

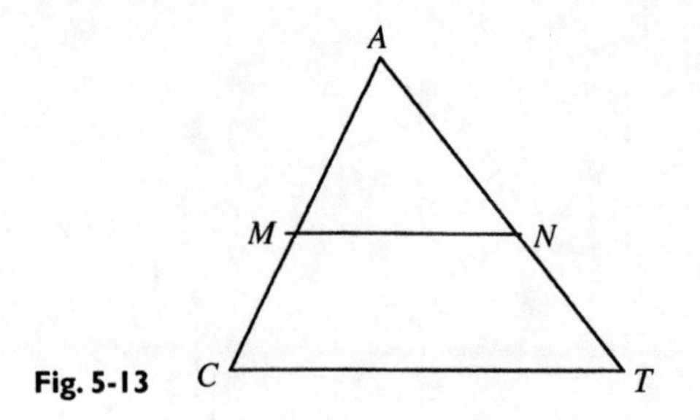

Fig. 5-13

11. In $\triangle CAT$, what is the measure of $\overline{MN}$ if it connects the midpoints of sides $\overline{AC}$ and $\overline{AT}$ and the measure of $\overline{CT}$ is 18?
12. In $\triangle CAT$, if $\overline{MN}$ connects the midpoints of sides $\overline{AC}$ and $\overline{AT}$ and the measure of $\overline{MN}$ is equal to 7, what is the measure of $\overline{CT}$?

Solutions to Supplementary Quadrilaterals Problems

1. The measure of the median of a trapezoid is equal to the average of the two base lengths. To solve for x, set up the equation as follows: $[(x + 2) + (4x - 3)]/2 = 7$. Now solve the equation for x.

$$5x - 1 = 14$$
$$5x = 15$$
$$x = 3$$

2. To determine the measures of the bases of the right trapezoid *TRAP,* simply substitute the value of x derived in Prob. 1. When 3 is substituted for x, the measures of the bases become 5 and 9.
3. The perimeter of *TRAP* is the sum of the sides. Add the measures of the bases derived in Prob. 2 to the measures of the sides given in Fig 5-11. The sum is as follows: $5 + 9 + 12 + 10 = 36$.
4. To calculate how much the median will increase if the value of x in the equation is changed, remember that the measure of the median is equal to the average of the two base lengths. Therefore, the measure of $\overline{MD}$ is equal to

$$\frac{(x + 2) + (4x - 3)}{2} = \frac{(7 + 2) + (4 \cdot 7 - 3)}{2} = \frac{9 + 25}{2} = 17$$

5. The measure of $\measuredangle TRA$ is 100°, because the sum of the interior angles of a quadrilateral is 360°.

6. The measure of diagonal $\overline{EL}$ is 14. The intersecting diagonals in a parallelogram bisect each other. If $\overline{EO}$ is the segment created when the diagonals of $\square SEAL$ intersect, then it is one-half the length of the diagonal.
7. The measure of $\overline{SO}$ is 5, because the line segment is created when the diagonals of $\square SEAL$ intersect. Therefore, $\overline{SO}$ is one-half the length of the diagonal $\overline{SA}$, which measures 10.
8. The measure of $\measuredangle ELA$ is equal to the measure of $\measuredangle SEL$, because they are alternating interior angles. $\overline{SE} \parallel \overline{LA}$ and $\overline{EL}$ is a transversal that crosses the parallel lines. So, the measure of $\measuredangle SEL = \measuredangle ELA = 30°$.
9. $\measuredangle EOA$ and $\measuredangle SOL$ are vertical angles, so they are equal in measure. Thus, $\measuredangle EOA = \measuredangle SOL = 75°$.
10. Opposite sides of a parallelogram are congruent, so the perimeter is equal to twice the measure of one side plus twice the measure of an adjacent side. $\overline{SL} = \overline{EA}$ and $\overline{SE} = \overline{LA}$. The sum of the four sides is 34.

$$34 = 2(6) + 2\,\overline{AL}$$
$$22 = 2\,\overline{AL}$$
$$11 = \overline{AL}$$

11. A line segment that joins the midpoints of two sides of a triangle is parallel to the third side and measures half the length of the third side. Therefore, the measure of $\overline{MN}$ is equal to one-half the measure of $\overline{CT}$. If the measure of $\overline{CT}$ is 18, then the measure of $\overline{MN}$ is 9.
12. A line segment that joins the midpoints of two sides of a triangle is parallel to the third side and measures half the length of the third side. Therefore, if the measure of $\overline{MN}$ is 7, then the measure of $\overline{CT}$ is twice that amount or 14.

Chapter 6

Inequalities in Geometry

The focus to this point has been upon showing that line segments, angles, triangles, and quadrilaterals are congruent. To do so, we have had to apply most of what is usually taught in algebra regarding the properties of equality, which were reviewed in Chap. 2 when we discussed deductive reasoning. In geometry, however, you will frequently encounter situations in which you must work with segments and angles that have unequal measure. In these situations, you will apply the properties of inequality that are taken from algebra.

Definitions to Know

Contrapositive. A statement that presents the complete opposite of a stated relationship and reverses the positions of the conditional (hypothesis) and the conclusion portions of the original statement. Thus, if the original statement is stated as "If *a,* then *b,*" the contrapositive is "If not *b,* then not *a.*"

Exterior angle of a triangle. An angle that is formed when one side of the triangle is extended, pictured in Fig. 6-1 as $\measuredangle WXZ$.

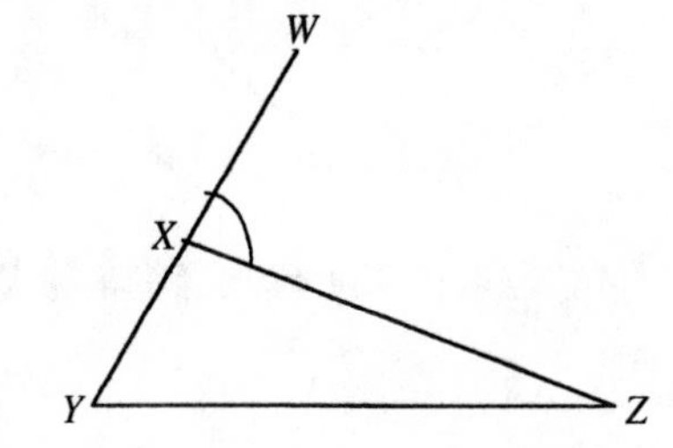

Fig. 6-1

Indirect proof. A proof that requires that you reason logically until you reach a contradiction of the hypothesis or some known fact. The indirect proof requires the following steps:

1. Assume temporarily that the conclusion is not true.
2. Reason logically until you reach a contradiction of an already established fact.
3. Point out that the temporary assumption must be false, and that the conclusion must then be true.

Inverse statement. A statement that presents the complete opposite of a stated relationship. Thus, if the original relationship is stated as "If *a*, then *b*," then the inverse statement is "If not *a*, then not *b*."

Logically equivalent statements. Two statements that are either both true or both false. Thus, a statement and its contrapositive are logically equivalent, but a statement is not logically equivalent to either its converse or its inverse.

Remote interior angles. Angles of a triangle that are at a distance from and not supplementary to a given exterior angle, identified in Fig. 6-1 as $\measuredangle Y$ and $\measuredangle Z$.

Venn diagram. A circle diagram that may be used to represent a conditional and its contrapositive. In Fig. 6-2, any point inside circle *a* is also inside circle *b*, so we can make the statement "If *a*, then *b*." At the same time, a point that is not inside circle *b* cannot be inside circle *a*, so the contrapositive "If not *b*, then not *a*" is also true.

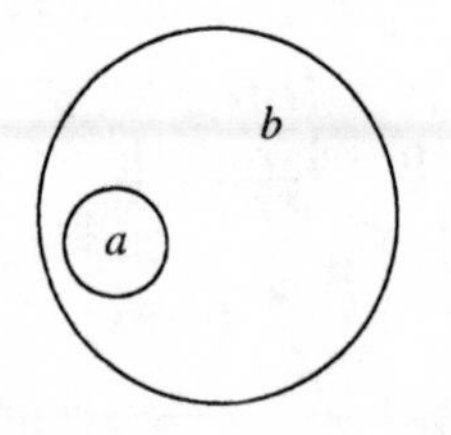

Fig. 6-2

Relevant Properties and Theorems

Property of Inequality I

If $a > b$ and $c \geq d$, then $a + c > b + d$.

Property of Inequality 2

If $a > b$ and $c > 0$, then $ac > bc$ and $a/c > b/c$.

Property of Inequality 3

If $a > b$ and $c < 0$, then $ac < bc$ and $a/c < b/c$.

Property of Inequality 4

If $a > b$ and $b > c$, then $a > c$.

Property of Inequality 5

If $a = b + c$ and $c > 0$, then $a > b$.

Theorem 1 (Exterior Inequality)

The measure of an exterior angle of a triangle is greater than the measure of either remote interior angle.

Theorem 2 (Side-Angle Relationship)

If one side of a triangle is longer than a second side, then the angle opposite the first side is larger than the angle opposite the second side.

Theorem 3 (Angle-Side Relationship)

If one angle of a triangle is larger than a second angle, then the side opposite the first angle is longer than the side opposite the second angle.

Theorem 4 (Perpendicular Segments)

The perpendicular segment from a point to a line or a plane is the shortest segment from the point to the line or the plane.

Theorem 5 (Triangle Inequality)

The sum of the lengths of any two sides of a triangle is greater than the length of the third side.

Theorem 6 (Side-Angle-Side Inequality)

If two sides of one triangle are congruent to two sides of another triangle, but the included angle of the first triangle is larger than the included angle of the second, then the third side of the first triangle is longer than the third side of the second triangle, as pictured in Fig. 6-3.

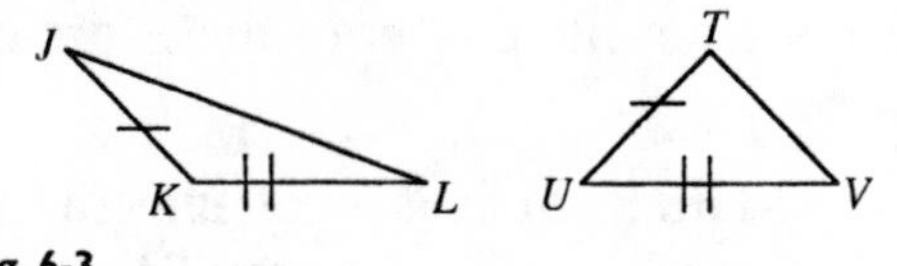

Fig. 6-3

Theorem 7 (Side-Side-Side Inequality)

If two sides of one triangle are congruent to two sides of another triangle, but the third side of the first triangle is longer than the third side of the second, then the included angle of the first triangle is larger than the included angle of the second, as pictured in Fig. 6-3.

EXAMPLE 1

Create a table in which you list (1) the statement, (2) the contrapositive, (3) the converse, and (4) the inverse of the following statement:

If $\overline{GO} = \overline{ON}$, then O is the midpoint of $\overline{GN}$.

Solution

	If ______, then ______.	**True/false**
Statement	If $\overline{GO} = \overline{ON}$, then O is the midpoint of $\overline{GN}$.	True
Contrapositive	If O is not the midpoint of $\overline{GN}$, then $\overline{GO} \neq \overline{ON}$.	False
Converse	If O is the midpoint of $\overline{GN}$, then $\overline{GO} = \overline{ON}$.	True
Inverse	If $\overline{GO} \neq \overline{ON}$, then O is not the midpoint of $\overline{GN}$.	True

EXAMPLE 2

Reword the following statement into conditional (if-then) form and illustrate the statement with a Venn diagram. Review each additional statement together with the given statement and decide what you can conclude. If you cannot conclude anything, then state "No conclusion."

Given: All Turnerville city employees are college graduates.

(*a*) Marian Jones is a college graduate.
(*b*) Jana Roan is a Turnerville city employee.

(*c*) Robert Wood is not a Turnerville city employee.
(*d*) Larry Taylor is not a college graduate.

Solution

Statement: If you are a city employee, then you must be a college graduate. (See Fig. 6-4.)

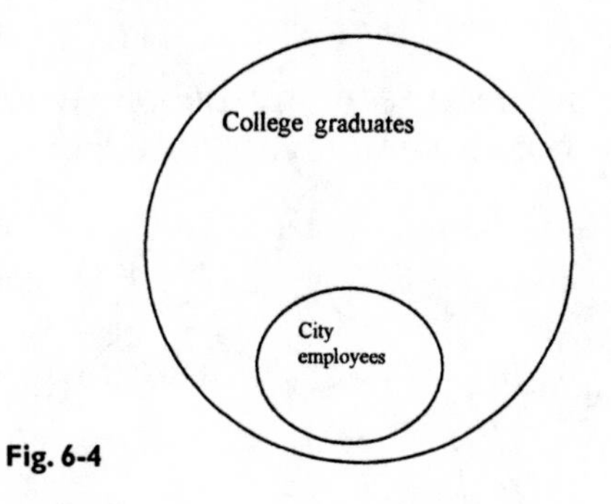

Fig. 6-4

(*a*) No conclusion can be established; she may or may not be a city employee. (See Fig. 6-5.)

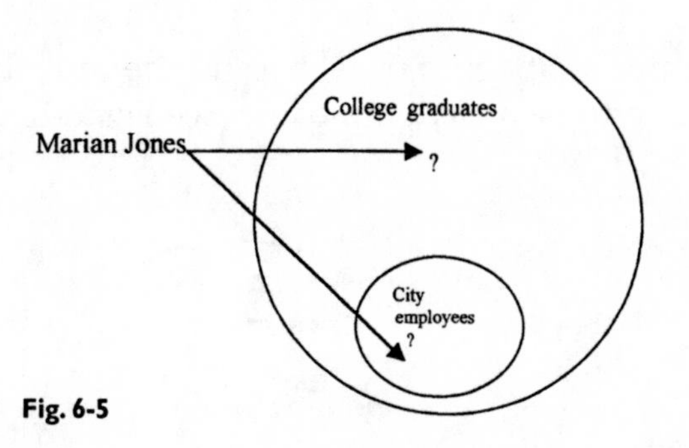

Fig. 6-5

(*b*) She must be a college graduate because she must be a college graduate to be a city employee. (See Fig. 6-6.)

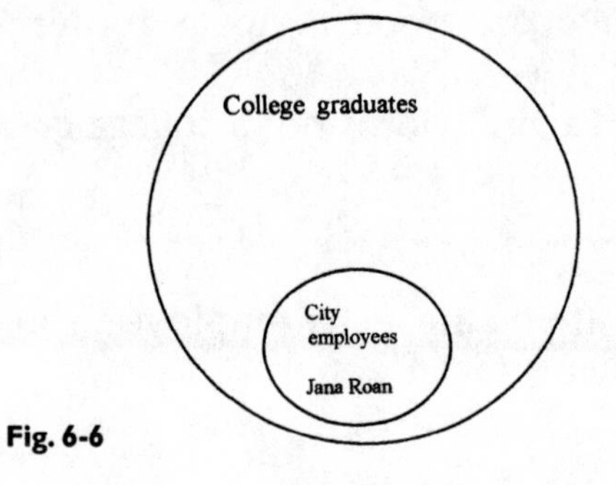

Fig. 6-6

(*c*) No conclusion, because many people who are college graduates are not city employees. (See Fig. 6-7.)

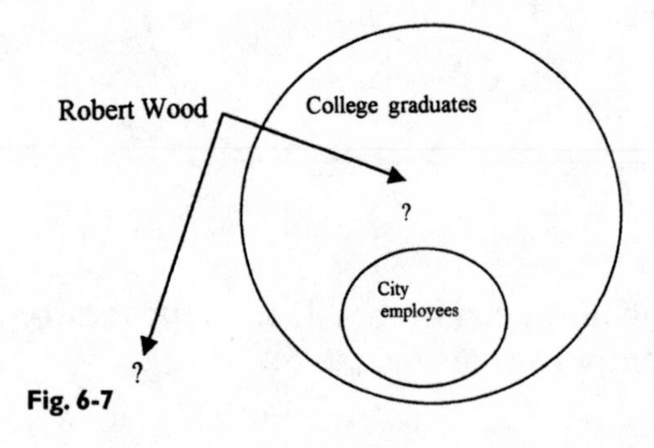

Fig. 6-7

(*d*) He is not a city employee, because he must be a college graduate to be a city employee. (See Fig. 6-8.)

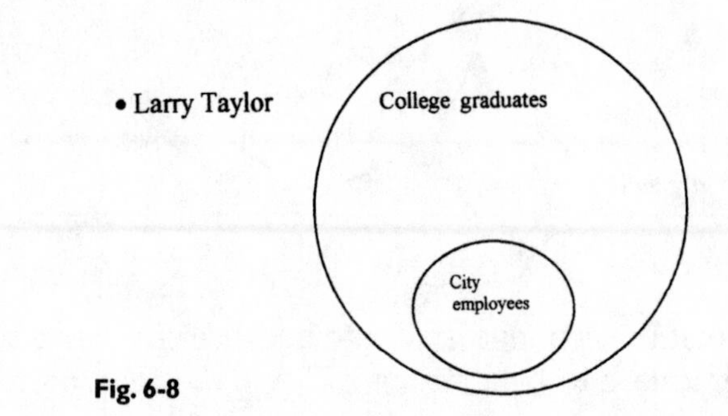

Fig. 6-8

EXAMPLE 3

If the lengths of two legs of a triangle are given as 4 and 8, what is the greatest measure of the third side? What is the least measure of the third side?

Solution

Let x equal the length of the third side. By applying the Triangle Inequality Theorem, we know that the sum of the lengths of any two sides of a triangle is greater than the length of the third side. Therefore, we can set up the equation as follows:

$$x + 4 > 8 \qquad 4 + 8 > x \qquad x + 8 > 4$$
$$x > 4 \qquad 12 > x \qquad x > 4$$

Thus, the length of the third side must be greater than 4 and less than 12.

EXAMPLE 4

Complete the statements by writing the correct symbol: $<$, $=$, or $>$. (See Fig. 6-9.)

$$\overline{QP}\,\_\_\_\_\_\,\overline{QR}$$
$$\overline{QW}\,\_\_\_\_\_\,12$$

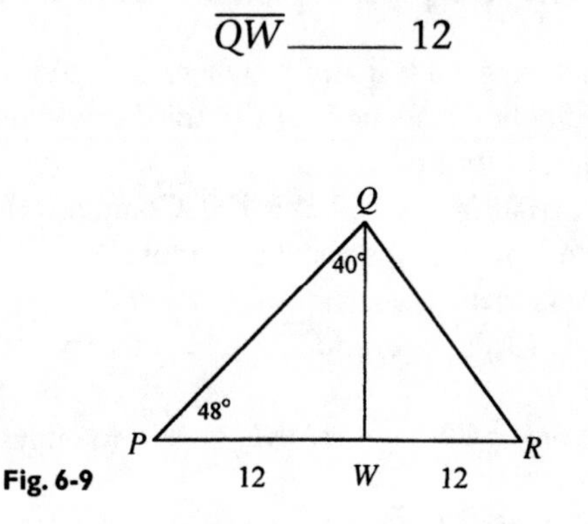

Fig. 6-9

Solution

$\overline{QP}$ is opposite $\measuredangle\overline{QWP}$, which is equal to $180° - (40° + 48°) = 180° - 88° = 92°$. $\overline{QR}$ is opposite $\measuredangle\overline{QWR}$, which is the supplement to $\measuredangle\overline{QWP}$. Therefore, $\measuredangle\overline{QWR} = 88°$.

We apply Theorem 4 (Angle-Side Relationship) that states if one angle of a triangle is larger than a second angle, then the side opposite the first angle is longer than the side opposite the second angle. Based on this theorem, we can complete the first statement as follows:

$$\overline{QP} > \overline{QR}$$

To determine the correct symbol to insert in the second statement, we must compute the measure of $\measuredangle WQR$ and compare it with the measure of $\measuredangle QRW$, which we know equals 48°. We also know that $\measuredangle QWR$ equals 88°. Thus, the measure of $\measuredangle WQR$ equals $180° - (48° + 88°) = 180° - 136° = 44°$.

We apply Theorem 4 (Angle-Side Relationship) that states if one angle of a triangle is larger than a second angle, then the side opposite the first angle is longer than the side opposite the second angle. Based on this theorem, we can complete the first statement as follows:

$$\overline{QW} > 12$$

Supplementary Inequalities Problems

1. The sides of a triangle are a and b (where $a > b$). Complete the following statement: The length of the third side must be greater than ______, but less than ______.
2. The sides of a triangle are 100 and 100. Complete the following statement: The length of the third side must be greater than ______, but less than ______.
3. In $\triangle DEF$, $\overline{DE} = 13$, $\overline{FD} = 8$, and $\overline{EF} = 11$. Identify the largest angle.
4. In $\triangle LMN$, $\measuredangle LMN = 70°$, and $\measuredangle MNL$, $= 60°$. Identify the longest segment.
5. In $\triangle MAX$ with sides of $\overline{MA} = 5$, $\overline{MX} = 7$, and $\overline{AX} = 6$, identify the largest angle and the smallest angle.
6. In $\triangle CAT$ with angles of $\measuredangle ATC = 65°$ and $\measuredangle TCA = 55°$, identify the longest and the shortest segments.

Refer to Fig. 6-10, and complete each statement by writing $<$, $=$, or $>$.

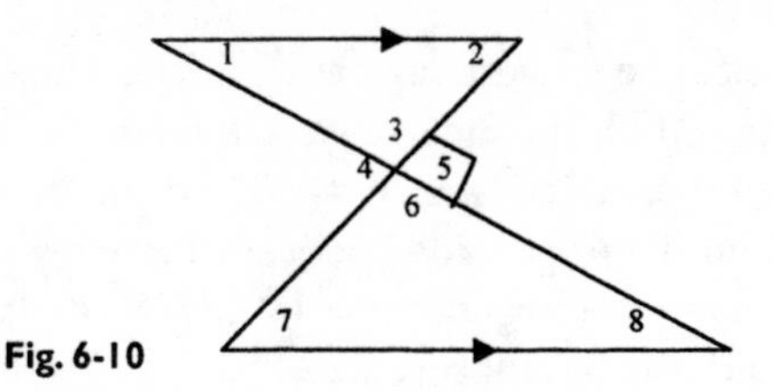

Fig. 6-10

7. $m\angle 1$ ______ $m\angle 3$
8. $m\angle 4$ ______ $m\angle 5$
9. $m\angle 6$ ______ $m\angle 8$
10. $m\angle 2$ ______ $m\angle 7$
11. $m\angle 3$ ______ $m\angle 6$
12. $m\angle 1$ ______ $m\angle 8$

Solutions to Supplementary Inequalities Problems

1. Let x equal the measure of the third side. By applying the Triangle Inequality Theorem, we know that the sum of the lengths of any two sides of a triangle is greater than the length of the third side.

$$x + a > b \qquad a + b > x \qquad x + b > a$$
$$x > b - a \qquad x < a + b \qquad x > a - b$$

Thus, the length of the third side must be greater than $b - a$ but less than $a + b$.

2. Let x equal the measure of the third side. By applying the Triangle Inequality Theorem, we know that the sum of the lengths of any two sides of a triangle is greater than the length of the third side.

$$x + 100 > 100 \qquad 100 + 100 > x \qquad x + 100 > 100$$
$$x > 0 \qquad 200 > x \qquad x > 0$$

The length of the third side must be greater than 0 and less than 200.

3. To solve this problem, sketch the triangle, placing the segments in correct relation to each other. When you do so, you find that

$\measuredangle DFE$ is opposite the longest segment $\overline{DE}$. Therefore, $\measuredangle DFE$ is the largest angle, based on the Side-Angle Theorem.

4. In $\triangle LMN$, the unidentified $\measuredangle NLM = 50°$, so the largest angle is $\measuredangle LMN$. Based on the Angle-Side Theorem, the longest segment in a triangle is opposite the largest angle. In $\triangle LMN$, the side opposite $\measuredangle LMN$ is $\overline{LN}$ and it is the longest side.

5. In $\triangle MAX$, the largest angle is opposite the longest side and the smallest angle is opposite the shortest side. Sketch the triangle, placing the identified segments in their respective positions. You will find that $\measuredangle MXA$ is opposite the shortest segment $\overline{MA}$ and $\measuredangle MAX$ is opposite the longest segment $\overline{MX}$.

6. In $\triangle CAT$, the longest side is opposite the largest angle and the shortest side is opposite the smallest angle. Sketch the triangle, carefully identifying the respective angles. You will find that the unidentified third angle $\measuredangle ACT = 180° - (55° + 65°) = 60°$. Therefore, $\measuredangle CAT$ is the smallest angle and $\overline{CT}$, the side opposite it, is, thus, the shortest segment. The largest angle is $\measuredangle CTA$, and $\overline{CA}$, the segment opposite the angle, is the largest segment.

7. $m\measuredangle 1 < m\measuredangle 3$. The vertical angles 3 and 6, 4 and 5 are each 90°. Because $\measuredangle 1 + \measuredangle 2 + \measuredangle 3 = 180°$, and $\measuredangle 3 = 90°$, the sum of angles 1 and 2 must equal 90°. Therefore, each angle must individually be less than 90°.

8. $m\measuredangle 4 = m\measuredangle 5$. Vertical angles are equal.

9. $m\measuredangle 6 > m\measuredangle 8$. The vertical angles 3 and 6, 4 and 5 are 90° each. Because $\measuredangle 6 + \measuredangle 7 + \measuredangle 8 = 180°$, and $\measuredangle 6 = 90°$, the sum of angles 7 and 8 must equal 90°. Therefore, each angle must individually be less than 90°.

10. $m\measuredangle 2 = m\measuredangle 7$. Interior angles formed when a transversal crosses parallel lines are equal in measure.

11. $m\measuredangle 3 = m\measuredangle 6$. Vertical angles are equal.

12. $m\measuredangle 1 = m\measuredangle 8$. Interior angles formed when a transversal crosses parallel lines are equal in measure.

Chapter 7

Similar Polygons

Read a map or engage in a hobby, and you will find that you are using the principles that we will review in this chapter. The legend on a map contains a scale that tells how many inches on the map represent a specific number of miles on the highways. Thus, you will usually see a scale similar to the one shown in Fig. 7-1. The map may also tell you how many kilometers to an inch (or some other measure), as we make greater use of the metric system.

Fig. 7-1 1 inch = 10 miles (approx.)

If you work with model airplanes or model cars, the principles in this chapter are also familiar to you, even if you have not identified them as being vital to building skills in geometry. Look at the box of a model airplane or car kit and you will see that the manufacturer identifies the way in which the finished model relates in size to the real item. You might work with a model identified as being in a 1:20 ratio with the original. This means that for every 1 inch (or 1 foot) of the model, the original measures 20 inches (or 20 feet). Naturally, a model that has a smaller difference between the two numbers is closer in size to the original. Thus, the model of a Stealth fighter jet that has a proportion of 1:100 probably contains less specific detail and appears less authentic than a model that relates to the original as 1:50. Both will be *similar* in appearance, but one will be large enough to permit greater use of details. Whatever the scale may be, the model represents the original in every way except size.

What does this have to do with geometry—and specifically with similar polygons? Quite a bit. The model of an airplane or car is a replica of the original, similar to the original in shape, details, and appearance. Every aspect of the model is directly related to the original in exactly the same proportion. Thus, if the model relates to the original as 1:50, one wing of the original is 50 times as large as the same wing in the same placement on the model; the fuselage of the original is 50 times as long and wide around as the fuselage of the model, and so on.

In geometry, we use this type of relationship to help us to solve for lengths and measures of polygons, using information that we have for one figure to determine lengths and measures of a similar figure.

Definitions to Know

Extended proportion. (See **proportion.**) Exists when three or more ratios are equal, as in the following: $e/f = g/h = j/k$ and $e{:}f = g{:}h = j{:}k$.

Extremes. The first and last terms of a proportion, as in the following in which the letters and numbers in bold print are the extremes: $\boldsymbol{m}{:}n = p{:}\boldsymbol{q}$; $\mathbf{5}{:}7 = 10{:}\mathbf{14}$.

Means of a proportion. The middle terms, as in the following in which n, p, 7, and 10, the letters and numbers not in bold print, are the means: $\boldsymbol{m}{:}n = p{:}\boldsymbol{q}$; $\mathbf{5}{:}7 = 10{:}\mathbf{14}$.

Means-extremes property of proportions. States that the product of the extremes of a proportion is equal to the product of the means. Thus, if $\boldsymbol{m}{:}\text{n} = p{:}\boldsymbol{q}$, then $\boldsymbol{mq} = np$; if $\mathbf{5}{:}7 = 10{:}\mathbf{14}$, then $\mathbf{5} \cdot \mathbf{14} = 7 \cdot 10$.

Proportion. An equation that states two ratios are equal, as you can see in the two equivalent forms of the same proportion: $g/h = j/k$ and $g{:}h = j{:}k$. Proportions have the following properties:

1. $a/b = c/d$ is the same as
   (a) $ad = bc$
   (b) $a/c = b/d$
   (c) $b/a = d/c$
   (d) $a + b/b = c + d/d$
2. If $a/b = c/d = e/f = \ldots$, then
   $a + c + e + \cdots/b + d + f + \cdots = a/b = \cdots$

Ratio. One number divided by another number. Thus, for all x and y, when $y \neq 0$, the ratio of x to y is x/y. The ratio is expressed in simplest form when both terms are reduced by factoring out their common factors.

Scale. A figure drawn to scale shows the correct shape of an object drawn in a convenient size.

Scale factor. For polygons, it is the ratio of the lengths of two corresponding sides of two similar polygons.

Similar polygons. Two polygons whose vertices can be paired so that the corresponding angles are congruent and whose corresponding sides have the same length ratio. To refer to similar polygons, list their corresponding vertices in the same order. Thus, if as in Fig. 7-2, polygon $ABCDE$ is similar to polygon $VWXYZ$, write polygon $ABCDE \sim$ polygon $VWXYZ$.

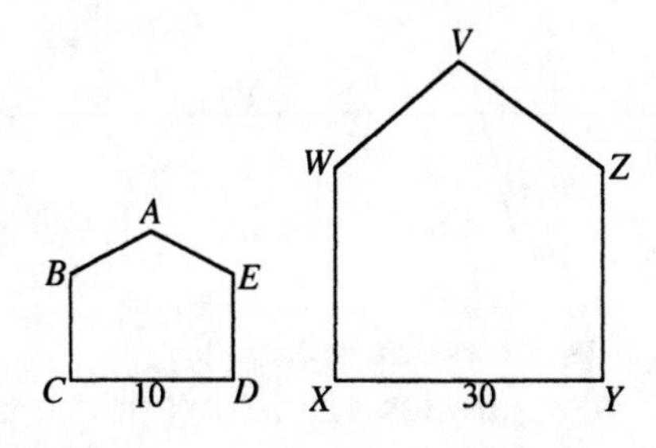

Fig. 7-2

If the two polygons are similar, we can also write the following:

$\measuredangle A \cong \measuredangle V \quad \measuredangle B \cong \measuredangle W \quad \measuredangle C \cong \measuredangle X \quad \measuredangle D \cong \measuredangle Y \quad \measuredangle E \cong \measuredangle Z$

and

$$\overline{AB}/\overline{VW} = \overline{BC}/\overline{WX} = \overline{CD}/\overline{XY} = \overline{DE}/\overline{YZ} = \overline{EA}/\overline{ZV}$$

Relevant Postulates and Theorems

Postulate 1 (Angle-Angle Similarity)

If two angles of one triangle are congruent to two angles of another triangle, as in Fig. 7-3, then the triangles are similar.

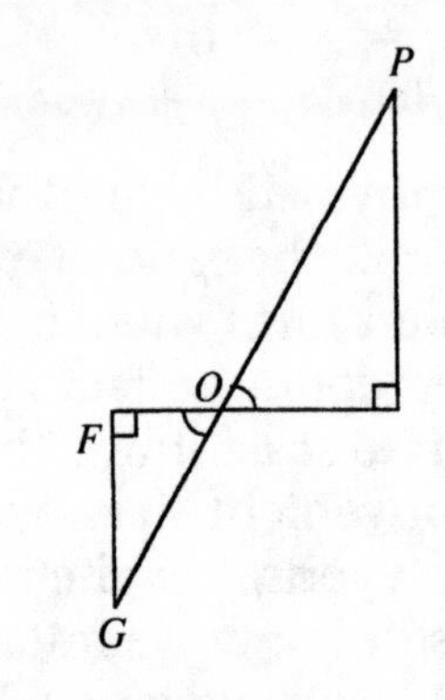

Fig. 7-3

Theorem 1 (Side-Angle-Side Similarity)

If an angle of one triangle is congruent to an angle of another triangle and the sides including those angles are in proportion as in Fig. 7-4, then the triangles are similar.

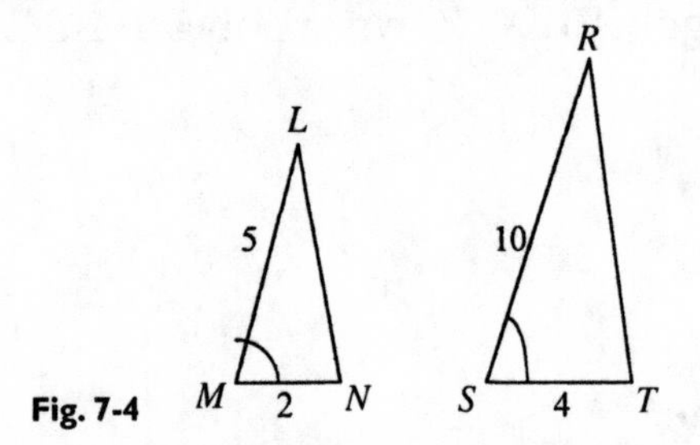

Fig. 7-4

Theorem 2 (Side-Side-Side Similarity)

If the sides of two triangles are in proportion as in Fig. 7-5, the triangles are similar.

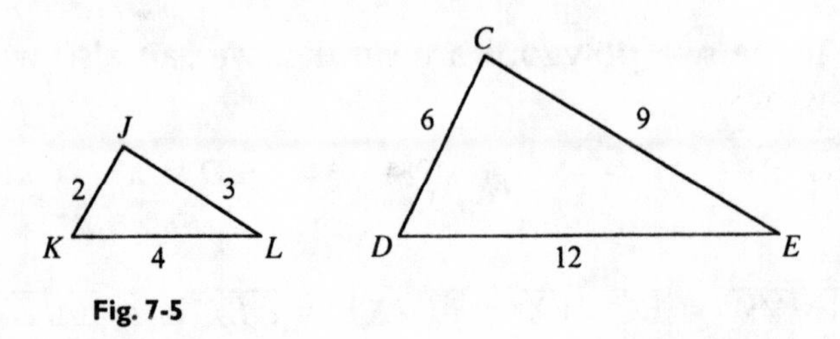

Fig. 7-5

Theorem 3 (Triangle Proportionality)

If a line parallel to one side of a triangle intersects the other two sides as in Fig. 7-6, then it divides those sides proportionally so that $\overline{GE}/\overline{GH} = \overline{GF}/\overline{GI}$.

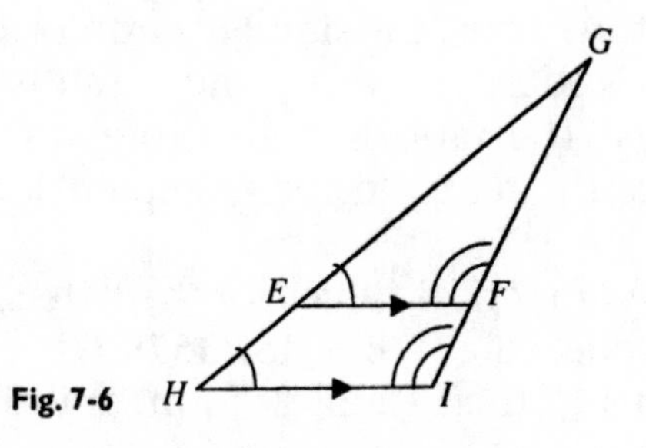

Fig. 7-6

Theorem 4 (Triangle Angle-Bisector)

If a ray bisects an angle of a triangle as in Fig. 7-7, then it divides the opposite side into segments proportional to the other two sides so that $\overline{BM}/\overline{BN} = \overline{LM}/\overline{LN}$.

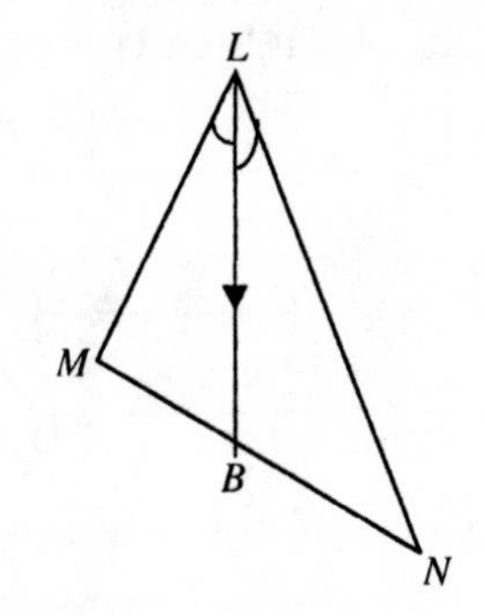

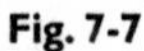

Fig. 7-7

EXAMPLE 1

In parts (*a*) through (*d*), $x = 18$, $y = 15$, and $z = 36$. Write each of the following ratios in simplest form:

(*a*) x to z
(*b*) z to y
(*c*) $x + y$ to z
(*d*) $z{:}y{:}x$

Solution

(*a*) By simply substituting the numerical values for the letters, the ratio can be expressed in the following manner: 18/36 or 18:36. However, to express the ratio in the simplest form, you must reduce both terms by factoring out their common factors. Thus, the simplest form of 18:36 is 1:2.

(*b*) Follow the procedure in part (*a*) and substitute the values for the letters. The ratio is then expressed as 36/15 or 36:15. Reduce the terms to their simplest form, and the answer is 12:5.

(*c*) Substitute the numerical values for the letters, and the ratio is then expressed as $(18 + 15)/36$ or $(18 + 15):36$. Reduce the terms to their simplest form, and the answer is $33:36 = 11:12$.

(*d*) Substitute the numerical values for the letters, and the ratio is then expressed as 36:15:18. Reduce the terms to their simplest form, and the answer is 12:5:6.

EXAMPLE 2

In Fig. 7-8, *FROG* is a parallelogram with dimensions as marked. Find the value of each of the following ratios:

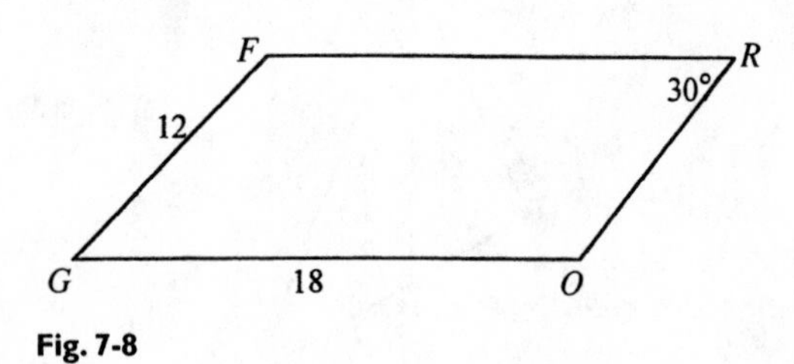

Fig. 7-8

(*a*) *FG*:*GO*
(*b*) $m\measuredangle F$:$m\measuredangle G$
(*c*) *RO*:perimeter of *FROG*

Solution

(*a*) Substitute the numerical values for the measures of the sides, and the ratio is then expressed as 12:18. Reduce the numbers by dividing by their greatest common factor, which is 6 in this case, and the ratio expressed in simplest terms is 2:3.

(*b*) Recall what you know about the angles in a parallelogram: (1) the sum of the angles in a parallelogram is 360° and (2) opposite angles in a parallelogram are equal in measure. Armed with this information, you can now determine the measures of angles *F* and *G:*

$$m\angle G = 30°$$

$$m\angle F = [360° - (2 \cdot 30°)]/2 = 150°$$

Substitute the measures in the ratio $m\angle F{:}m\angle G = 150{:}30$. To express the ratio in simplest terms, divide both measures by the greatest common factor of each, which is 30 in this case, and the ratio is expressed in simplest terms as 5:1.

(*c*) Recall that the opposite sides of a parallelogram are equal in length and that the perimeter of the parallelogram is the sum of the measures of the four sides. $\overline{RO}$ is opposite $\overline{FG}$, a side with a length of 12. The perimeter of □*FROG* is equal to $12 + 12 + 18 + 18 = 60$. Thus, the ratio requested is 12:60. To express the ratio in simplest terms, divide both values by their greatest common factor, which is 12 in this case, the simplest form of the ratio is 1:5.

EXAMPLE 3

Find the value of *t* in parts (*a*) through (*d*).

(*a*) $t/4 = 5/2$
(*b*) $5/2 = 7/3t$
(*c*) $7/(6t - 4) = 9/(4t + 6)$
(*d*) $(t + 5)/4 = 1/2$

Solution

(*a*) Using the definition of the means-extremes property of proportions, you know that the product of the extremes equals the product of the means. The ratio $t/4 = 5/2$ can be rewritten as $t{:}4 = 5{:}2$, in which the extremes are *t* and 2, and the means are 4 and 5. Thus,

$$t \cdot 2 = 4 \cdot 5$$

$$2t = 20$$

$$t = 10$$

(*b*) Using the definition of the means-extremes property of proportions, you know that the product of the extremes equals the product of the means. The ratio $5/2 = 7/3t$ can be rewritten as $5:2 = 7:3t$, in which the extremes are 5 and $3t$, and the means are 2 and 7. Thus,

$$5 \cdot 3t = 2 \cdot 7$$
$$15t = 14$$
$$t = 14/15$$

(*c*) Using the definition of the means-extremes property of proportions, you know that the product of the extremes equals the product of the means. The ratio $7/(6t - 4) = 9/(4t + 6)$ can be rewritten as $7:(6t - 4) = 9:(4t + 6)$, in which the extremes are 7 and $(4t + 6)$ and the means are $(6t - 4)$ and 9. Thus,

$$7 \cdot (4t + 6) = 9 \cdot (6t - 4)$$
$$28t + 42 = 54t - 36$$
$$42 = 26t - 36$$
$$78 = 26t$$
$$78/26 = t$$
$$3 = t$$

(*d*) Using the definition of the means-extremes property of proportions, you know that the product of the extremes equals the product of the means. The ratio $(t + 5)/4 = 1/2$ can be rewritten as $(t + 5):4 = 1:2$, in which the extremes are $(t + 5)$ and 2 and the means are 4 and 1. Thus,

$$(t + 5) \cdot 2 = 4 \cdot 1$$
$$2t + 10 = 4$$
$$2t = -6$$
$$t = -3$$

Supplementary Similar Polygons Problems

1. Given that two sides of a rectangle have lengths of 40 and 64, find in simplest form the ratio of

(a) The length of the longer side to the shorter side
(b) The length of the shorter side to the perimeter
(c) The measure of any angle of the rectangle to the sum of the measures of the interior angles of the rectangle.

2. If the ratio of three variables $a:b:c = 3:4:5$, and $c = 30$, then $a =$ ____ and $b =$ ____.
3. If $6/51 = d/17$, then $d =$ ____.
4. If $x/y = z/13$, then $(x + y)/$____ $=$ ____$/13$.

Refer to the two similar polygons in Fig. 7-9 to complete Probs. 5 through 8.

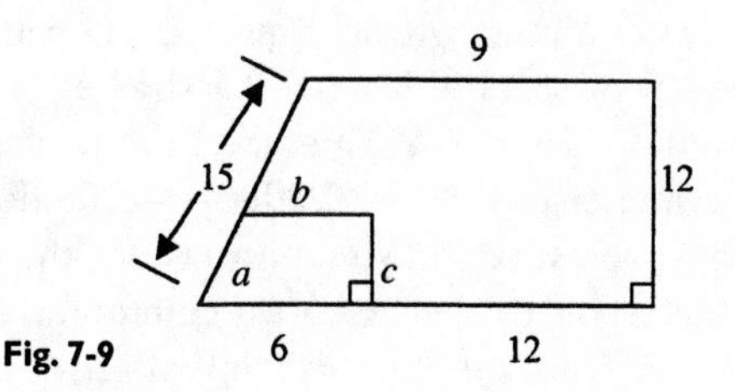

Fig. 7-9

5. What is the value of *a*?
6. What is the value of *b*?
7. What is the ratio of *c* and the perimeter of the larger polygon?
8. What is the ratio of *b* and the perimeter of the smaller polygon?

Refer to Fig. 7-10 to answer Probs. 9 and 10.

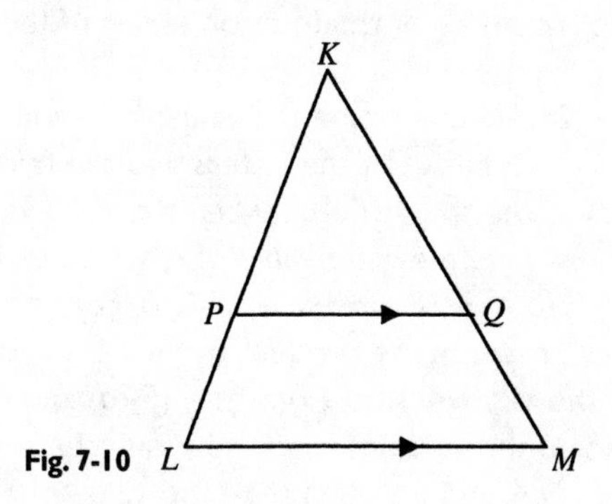

Fig. 7-10

9. If $\overline{KQ} = 14$, $\overline{QM} = 6$, and $\overline{PL} = 9$, then what is the measure of $\overline{KP}$?
10. If $\overline{KP} = 12$, $\overline{PL} = 6$, and $\overline{PQ} = 16$, what is the measure of $\overline{LM}$?

Solutions to Supplementary Similar Polygons Problems

1. (*a*) 8:5. The ratio of the longer side to the shorter side of a rectangle with sides that have lengths of 40 and 64 is 64:40. To provide the answer in simplest terms, you must reduce the terms by dividing both terms by their greatest common factor, which in this case is 8. Thus, the ratio in simplest terms of the longer side to the shorter side is 8:5.

   (*b*) 5:26. To calculate the ratio of the shorter side to the perimeter, you must first calculate the measure of the perimeter P of the rectangle. You know from previous chapters that opposite sides of a rectangle are equal, so the perimeter P of a rectangle having sides of length 40 and 64 equals $2(40) + 2(64) = 80 + 128 = 208$. Thus, the ratio of the shorter side to the perimeter is $40/P = 40/208 = 40{:}208$. To provide the answer in simplest terms, you must reduce the terms by dividing both terms by their greatest common factor, which in this case is 8. Thus, the ratio in simplest terms of the longer side to the shorter side is 5:26.

   (*c*) 1:4. In any rectangle, the sum of the interior angles is 360° and all four angles are right angles with a measure of 90° each. Thus, any one angle may be placed in proportion to the sum: 90:360. To provide the answer in simplest terms, you must reduce the terms by dividing both terms by their greatest common factor, which in this case is 90. Thus, the ratio in simplest terms of any angle of a rectangle to the sum of the measures of the interior angles of the rectangle is 1:4.

2. $a = 18$ and $b = 24$. To determine the values of a and b when $c = 30$, set up a ratio between the letters and the numbers. To find the value of a, set up a ratio between the ratio value of a, which is 3, and a versus the ratio value of c, which is 5, and the value of c. Thus, $3/a = 5/30$, or $3{:}a = 5{:}30$. According to the means-extremes property, the product of the means is equal to the product of the extremes, so $(3)(30) = (5)(a)$ and $18 = a$.

   To determine the value of b, set up a ratio between the ratio value of b, which is 4, and b versus the ratio value of c and the value of c. Thus, $4/b = 5/30$, or $4{:}b = 5{:}30$. Applying once again the means-extremes property, $(4)(30) = (5)(b)$, and $24 = b$.

3. $d = 2$. In the ratio $6/51 = d/17$, 6 and 17 are the extremes and d

and 51 are the means. Applying the means-extremes property, $(6)(17) = (51)(d)$, and $d = 2$.

4. $(x + y)/y = (z + 13)/13$. Use the definition of proportion to determine the specific property that applies to the ratio equation.
5. $a = 5$. If two polygons are similar, then their corresponding vertices are congruent and the lengths of their corresponding sides have the same ratio. Thus, $a{:}15 = 6{:}18$, or $a/15 = 6/18$. Multiply the means and the extremes and set them equal to each other:

$$(a)(18) = (6)(15)$$
$$a = 5$$

6. $b = 3$. If two polygons are similar, then their corresponding vertices are congruent and the lengths of their corresponding sides have the same ratio. Thus, $b{:}9 = 6{:}18$, or $b/9 = 6/18$. Multiply the means and the extremes and set them equal to each other:

$$(b)(18) = (6)(9)$$
$$b = 3$$

7. 2:27. To set up the ratio, you must first find the values of c and the perimeter of the larger polygon. First set up the ratio as you did in finding a and b. Thus, $c{:}12 = 6{:}18$, or $c/12 = 6/18$. Multiply the means and the extremes and set them equal to each other:

$$(c)(18) = (6)(12)$$
$$c = 4$$

The measure of the perimeter (P) of the larger polygon is equal to the sum of the lengths of the sides: $9 + 15 + 18 + 12 = 54 = P$. The ratio of c to P is expressed as $c{:}P$ or 4:54. To express the ratio in simplest terms, divide both terms by their greatest common factor, which is 2 in this case. Thus, the ratio of c to the perimeter of the larger polygon expressed in simplest terms is 2:27.

8. 1:6. To set up the ratio, you must use 3, the value for b already calculated, and find the perimeter of the smaller polygon. The measure of the perimeter (p) of the smaller polygon is equal to the sum of the lengths of the sides: $5 + 3 + 4 + 6 = 18 = p$. The ratio of b to p is expressed as $b{:}p = 3{:}18$. To express the ratio in simplest terms, divide both terms by their greatest common factor, which is 3 in this case. Thus, the ratio of b to the perimeter of the smaller polygon expressed in simplest terms is 1:6.

9. $KP = 21$. The Triangle Proportionality Theorem (Theorem 3) states that a line parallel to one side of a triangle that intersects the other two sides also divides those sides proportionally. Thus, $\overline{KQ}:\overline{QM} = \overline{KP}:\overline{PL}$, or $\overline{KQ}/\overline{QM} = \overline{KP}/\overline{PL}$. When we substitute the values, $14:6 = \overline{KP}:9$. To solve for $\overline{KP}$, set the product of the extremes equal to the products of the means:

$$(14)(9) = (\overline{KP})(6)$$
$$21 = \overline{KP}$$

10. $\overline{LM} = 24$. Once again, apply the Triangle Proportionality Theorem (Theorem 3). Thus, $\overline{KP}:\overline{KL} = \overline{PQ}:\overline{LM}$, or $\overline{KP}/\overline{KL} = \overline{PQ}/\overline{LM}$. When we substitute the values, $12:18 = 16:\overline{LM}$. To solve for $\overline{LM}$, set the product of the extremes equal to the product of the means:

$$(12)(\overline{LM}) = (18)(16)$$
$$\overline{LM} = 24$$

Chapter 8

Right Triangles

Many of the skills related to polygons are valuable in working with right triangles, but this chapter will not duplicate earlier discussions. Instead, we will examine such concepts as determining the geometric mean between two numbers, applying the Pythagorean Theorem, and simplifying numbers under the radical sign, skills which are more specific to right triangles. In addition, knowing the special relationships that exist between certain angle measures and the sides in right triangles will allow you to more easily identify the individual and overall dimensions of the many right triangles that we encounter in daily life.

Definitions to Know

Altitude of a triangle. A perpendicular segment drawn from a vertex to the side opposite the vertex. Its length is the geometric mean between the segments of the hypotenuse.

Geometric mean. The middle variable x between a and b in the proportion $a/x = x/b$, for positive numbers a, b, and x.

Hypotenuse. In a right triangle, it is the side opposite the right angle.

Legs. The sides of the triangle not opposite the right angle.

Pythagorean triple. Any triple of positive integers a, b, and c such that $a^2 + b^2 = c^2$, where a, b, and c are the lengths of the sides of a right triangle.

Radical in simplest form. A radical having 1 as the only perfect square under the radical sign, no fraction under the radical sign, and no fraction with a radical in its denominator.

Relevant Theorems

Theorem 1

An altitude drawn to the hypotenuse of a right triangle, as in Fig. 8-1, creates two triangles that are similar to each other and to the original triangle.

Corollary to Theorem 1

When the altitude is drawn to the hypotenuse of a right triangle, the length of the altitude is the geometric mean between the segments of the hypotenuse, and each leg is the geometric mean between the hypotenuse and the segment of the hypotenuse that is adjacent to that leg.

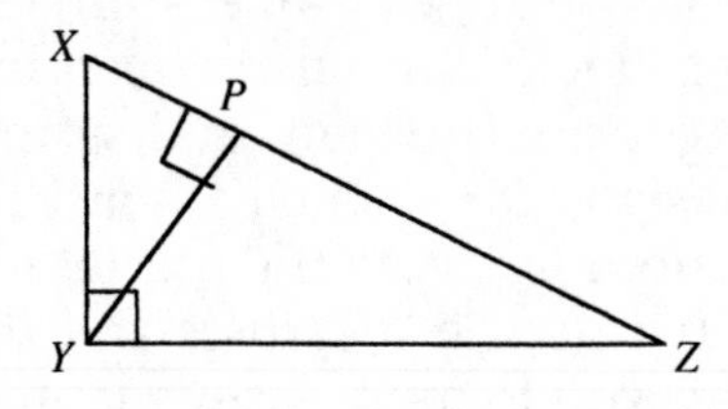

Fig. 8-1

Theorem 2 (Pythagorean Theorem)

In a right triangle, as depicted in Fig. 8-2, the square of the hypotenuse is equal to the sum of the squares of the legs, or $c^2 = a^2 + b^2$ and the $m\measuredangle C = 90°$.

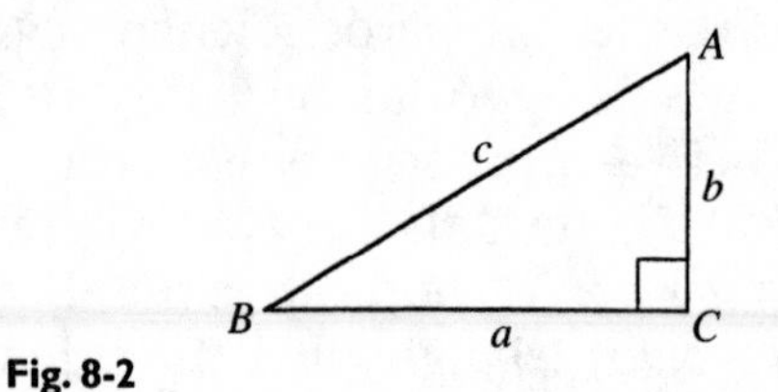

Fig. 8-2

Theorem 3

If the square of one side of a triangle is less than the sum of the squares of the other two sides, $c^2 < a^2 + b^2$, then $m\measuredangle C < 90°$ and the triangle is an acute triangle.

Theorem 4

If the square of one side of a triangle is greater than the sum of the squares of the other two sides, $c^2 > a^2 + b^2$, then $m\measuredangle C > 90°$ and the triangle is an obtuse triangle.

Theorem 5 (45°-45°-90° Triangle)
In a triangle with angle measures of 45°, 45°, and 90°, as in Fig. 8-3, the hypotenuse is $\sqrt{2}$ times as long as a leg.

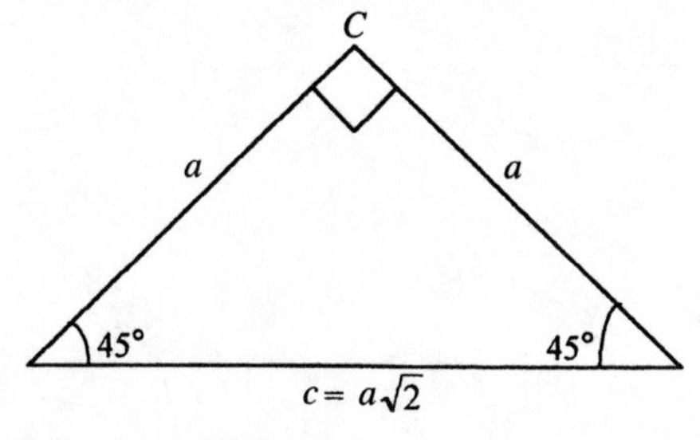

Fig. 8-3

Theorem 6 (30°-60°-90°)
In a triangle with angle measures of 30°, 60°, and 90°, as in Fig. 8-4, the hypotenuse is twice as long as the shorter leg and the longer leg is $\sqrt{3}$ times as long as the shorter leg.

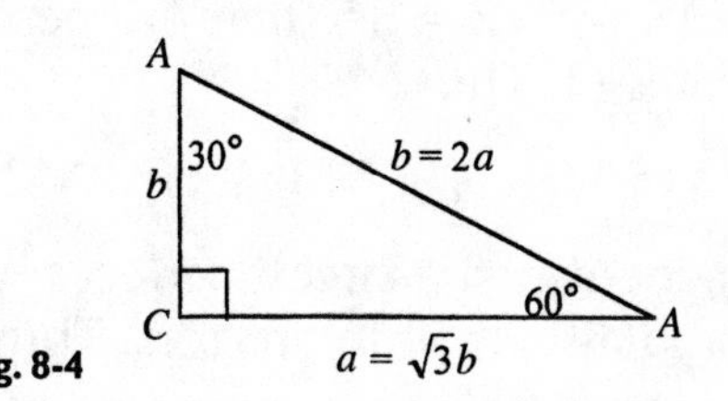

Fig. 8-4

The unique nature of the right triangle has led to the identification of specific side lengths that are common to right triangles, and these appear in the chart below. The numbers in bold print are the basic unit combinations that satisfy the equation of $c^2 = a^2 + b^2$, and the numbers listed below each set in bold print offer multiples of the original numbers that may be extended as needed to identify right triangles.

Right Triangle Side Lengths

3, 4, 5	5, 12, 13	8, 15, 17	7, 24, 25
6, 8, 10	**10, 24, 26**	**16, 30, 34**	**14, 48, 50**
9, 12, 15	**15, 36, 39**	**24, 45, 51**	**21, 72, 75**
12, 16, 20	**20, 48, 52**	**32, 60, 68**	**28, 96, 100**
15, 20, 25	**25, 60, 65**	**40, 75, 85**	**35, 120, 125**

EXAMPLE 1

Place the following radicals into simplest form:

(*a*) $\sqrt{18}$
(*b*) $\sqrt{500}$
(*c*) $4\sqrt{32}$
(*d*) $\dfrac{12}{\sqrt{3}}$
(*e*) $\dfrac{\sqrt{3}}{\sqrt{6}}$
(*f*) $\sqrt{\dfrac{3}{11}}$
(*g*) $\sqrt{7} \cdot \sqrt{63}$

Solution

(*a*) Divide the number under the radical sign into the largest perfect square that you can find and any remaining factors. Because $18 = 9 \cdot 2 = 3^2$, there is a perfect square under the radical sign. Thus,

$$\sqrt{18} = \sqrt{9 \cdot 2} = \sqrt{3^2 \cdot 2} = 3\sqrt{2}$$

(*b*) As in part (*a*), identify the largest perfect square under the radical sign and all remaining factors. Thus,

$$\sqrt{500} = \sqrt{25 \cdot 20} = \sqrt{5 \cdot 5 \cdot 4 \cdot 5} = \sqrt{5 \cdot 5}(\sqrt{4})(\sqrt{5})$$
$$= 5 \cdot 2 \cdot \sqrt{5} = 10\sqrt{5}$$

(*c*) Identify the largest perfect square under the radical sign and all remaining factors. Thus,

$$4\sqrt{32} = 4\sqrt{4 \cdot 4 \cdot 2} = 4 \cdot 4\sqrt{2} = 16\sqrt{2}$$

(*d*) The radical cannot remain in the denominator if the number is to be placed in simplest terms. To modify the equation, you must multiply the denominator by itself to remove the 3 from under the radical sign. Be careful, however, to multiply the numerator by anything by which you multipy the denominator. Thus,

$$\frac{12}{\sqrt{3}} = \frac{12}{\sqrt{3}} \cdot \frac{\sqrt{3}}{\sqrt{3}} = \frac{12\sqrt{3}}{3}$$

Now that you have eliminated the radical in the denominator, simplify the fraction by dividing the numerator and denominator by 3. The answer is $4\sqrt{3}$.

(*e*) The radical cannot remain in the denominator if the number is to be placed in simplest terms. To modify the equation, you must multiply the denominator by itself to remove the 6 from under the radical sign. Be careful, however, to multiply the numerator by anything by which you multipy the denominator. Thus,

$$\frac{\sqrt{3}}{\sqrt{6}} = \frac{\sqrt{3}}{\sqrt{6}} \cdot \frac{\sqrt{6}}{\sqrt{6}} = \frac{\sqrt{3 \cdot 6}}{6} = \frac{\sqrt{3 \cdot 3 \cdot 2}}{6} = \frac{3\sqrt{2}}{6} = \frac{\sqrt{2}}{2}$$

Reduce the fraction to lowest terms to derive the final answer of $\sqrt{2}/2$.

(*f*) The fraction must be removed from under the radical sign. First place the numbers under the radical sign into two radical signs, then multiply the numerator and the denominator by $\sqrt{11}$ to remove the radical sign from the denominator. Thus,

$$\frac{3}{\sqrt{11}} = \frac{\sqrt{3}}{\sqrt{11}} \cdot \frac{\sqrt{11}}{\sqrt{11}} = \frac{\sqrt{3 \cdot 11}}{11} = \frac{\sqrt{33}}{11}$$

The number 33 cannot be reduced any further.

(*g*) Break the numbers into their factors and identify any perfect squares that you can take the square root of and remove them from under the radical sign.

$$\sqrt{7} \cdot \sqrt{7 \cdot 9} = \sqrt{7 \cdot 7 \cdot 3 \cdot 3} = 7 \cdot 3 = 21$$

EXAMPLE 2

Given two positive numbers a and b, identify the geometric mean x between the two numbers identified by solving the proportion.

(*a*) 8 and 12
(*b*) 5 and 9
(*c*) 6 and 9

Solution

(*a*) The proportion used to identify the geometric mean is $a/x = x/b$. Substituting the values in the proportion, you have $8/x = x/12$. Now, solve for x. Recall from Chap. 7 that the product of the extremes is equal to the product of the means. Thus,

$$(8)(12) = x \cdot x$$
$$4 \cdot 4 \cdot 2 \cdot 3 = x^2$$
$$\sqrt{4 \cdot 4 \cdot 2 \cdot 3} = x$$
$$4\sqrt{6} = x \quad \text{(the geometric mean between 8 and 12)}$$

(*b*) The proportion used to identify the geometric mean is $a/x = x/b$. Substituting the values in the proportion, you have $5/x = x/9$. Now, solve for x. Recall from Chap. 7 that the product of the extremes is equal to the product of the means. Thus,

$$(5)(9) = x^2$$
$$5 \cdot 3 \cdot 3 = x^2$$
$$3\sqrt{5} = x \quad \text{(the geometric mean between 5 and 9)}$$

(*c*) The proportion used to identify the geometric mean is $a/x = x/b$. Substituting the values in the proportion, you have $6/x = x/9$. Now, solve for x. Recall from Chap. 7 that the product of the extremes is equal to the product of the means. Thus,

$$(6)(9) = x^2$$
$$\sqrt{6 \cdot 9} = \sqrt{3 \cdot 2 \cdot 3 \cdot 3} = x$$
$$3\sqrt{6} = x \quad \text{(the geometric mean between 6 and 9)}$$

EXAMPLE 3

Refer to Fig. 8-5 to complete parts (*a*) through (*d*).

(*a*) If $\overline{HQ} = 7$ and $\overline{QI} = 10$, find $\overline{GQ}$.
(*b*) If $\overline{HQ} = 4$ and $\overline{QI} = 7$, find $\overline{GQ}$.
(*c*) If $\overline{GQ} = 10$ and $\overline{HQ} = 3$, find $\overline{QI}$.
(*d*) If $\overline{GQ} = 5$ and $\overline{QI} = 9$, find $\overline{HQ}$.

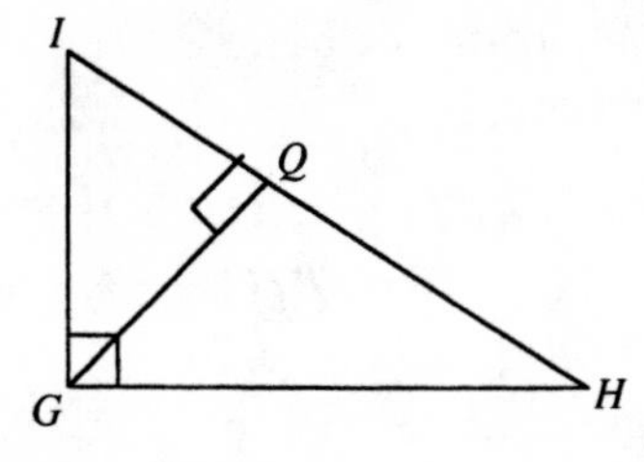

Fig. 8-5

Solution

(*a*) In $\triangle GHI$, $\overline{HI}$ is the hypotenuse situated opposite the right angle G. $\overline{GQ}$ is the altitude drawn to the hypotenuse. Apply the Corollary to Theorem 1 that states that the altitude drawn to the hypotenuse of a right triangle is the geometric mean between the segments of the hypotenuse. Thus,

$$\frac{\overline{HQ}}{\overline{GQ}} = \frac{\overline{GQ}}{\overline{QI}}$$

$$\frac{7}{\overline{GQ}} = \frac{\overline{GQ}}{10}$$

Recall that the product of the extremes equals the product of the means. So,

$$(7)(10) = (\overline{GQ})(\overline{GQ})$$

$$70 = \overline{GQ}^2$$

$$\sqrt{70} = \overline{GQ}$$

(*b*) In $\triangle GHI$, $\overline{HI}$ is the hypotenuse situated opposite the right angle G. $\overline{GQ}$ is the altitude drawn to the hypotenuse. Apply the Corollary to Theorem 1 that states that the altitude drawn to the hypotenuse of a right triangle is the geometric mean between the segments of the hypotenuse. Thus,

$$\frac{\overline{HQ}}{\overline{GQ}} = \frac{\overline{GQ}}{\overline{QI}}$$

$$\frac{4}{\overline{GQ}} = \frac{\overline{GQ}}{7}$$

Recall that the product of the extremes equals the product of the extremes.

$$(4)(7) = (\overline{GQ})(\overline{GQ})$$
$$\sqrt{4 \cdot 7} = \overline{GQ}^2$$
$$2\sqrt{7} = \overline{GQ}$$

(*c*) In $\triangle GHI$, $\overline{HI}$ is the hypotenuse situated opposite the right angle G. $\overline{GQ}$ is the altitude drawn to the hypotenuse. Apply the Corollary to Theorem 1 that states that the altitude drawn to the hypotenuse of a right triangle is the geometric mean between the segments of the hypotenuse. Thus,

$$\frac{\overline{HQ}}{\overline{GQ}} = \frac{\overline{GQ}}{\overline{QI}}$$
$$\frac{3}{10} = \frac{10}{\overline{QI}}$$

Recall that the product of the extremes equals the product of the extremes.

$$(3)(\overline{QI}) = (10)(10)$$
$$\overline{QI} = \frac{100}{3}$$

(*d*) In $\triangle GHI$, $\overline{HI}$ is the hypotenuse situated opposite the right angle G. $\overline{GQ}$ is the altitude drawn to the hypotenuse. Apply the Corollary to Theorem 1 that states that the altitude drawn to the hypotenuse of a right triangle is the geometric mean between the segments of the hypotenuse. Thus,

$$\frac{\overline{HQ}}{\overline{GQ}} = \frac{\overline{GQ}}{\overline{QI}}$$
$$\frac{\overline{HQ}}{5} = \frac{5}{9}$$

Recall that the product of the extremes equals the product of the extremes.

$$\overline{HQ}\,(9) = (5)(5)$$

$$\overline{HQ} = \frac{25}{9}$$

EXAMPLE 4

A triangle contains a right angle and two sides measuring a and b and a hypotenuse c, as in Fig. 8-6. What is the value of c when $a = 8$ and $b = 15$?

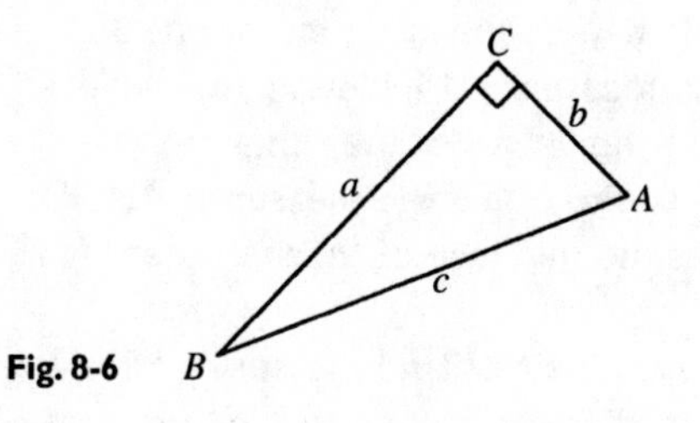

Fig. 8-6

Solution

The triangle contains a right angle, so you can apply the Pythagorean Theorem $c^2 = a^2 + b^2$ to calculate the lengths of the sides and hypotenuse. Substitute the values in the equation. Thus,

$$c^2 = 8^2 + 15^2 = 64 + 225 = 289$$

$$c = 17$$

Supplementary Right Triangles Problems

1. What is the geometric mean between 7 and 11? 4 and 5? 13 and 16?

Refer to Fig. 8-7 to complete Probs. 2 through 4.

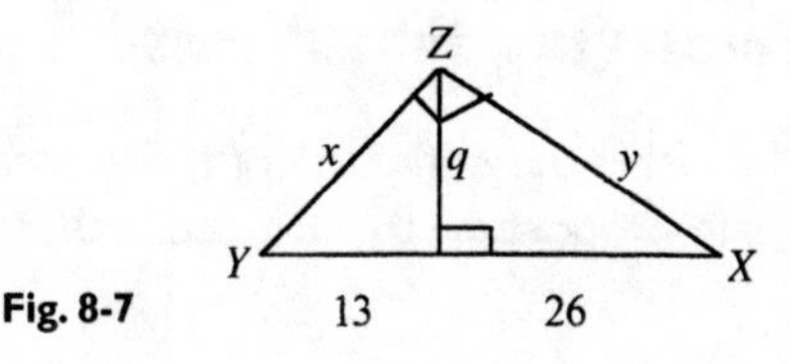

Fig. 8-7

2. Find the value of q, an altitude drawn from the right angle Z to intersect the hypotenuse $\overline{XY}$.
3. Find the length of side y, the side opposite $\measuredangle Y$ in the larger triangle in the figure.
4. Find the length of side x, the side opposite $\measuredangle X$ in the larger triangle in the figure.
5. What is the length of the diagonal of a rectangle with sides of 12 and 16?
6. If the legs of a right triangle are 11 and 13, what is the length of the hypotenuse?
7. In a right triangle having one angle measuring 30° and the leg opposite that angle measuring 14, identify (*a*) the length of the hypotenuse and (*b*) the length of the other leg.
8. In a right triangle with one angle measuring 45°, if one leg measures 25, what is the measure of (*a*) the other leg and (*b*) the hypotenuse?
9. City planners have a plot of land measuring $85\sqrt{2}$ by $85\sqrt{2}$ feet that they want to divide in half so that two groups of children will have exactly the same area in which to play. The planners also want to create a running track of maximum length, which will divide the land and that both groups will share. (*a*) Where will they divide the land? (*b*) What will be the length of the running track?
10. If $\triangle EFG$ in Fig. 8-8 is an equilateral triangle, what is the measure of the altitude a?

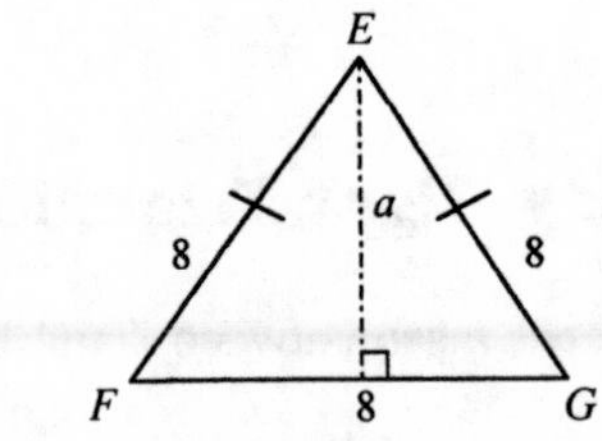

Fig. 8-8

Solutions to Supplementary Right Triangles Problems

1. $\sqrt{77}, 2\sqrt{5}, 4\sqrt{13}$. Using the definition of the geometric mean, set the given numbers in proportion to each other, then solve for the variable:

$$\frac{7}{x} = \frac{x}{11} \qquad \frac{4}{x} = \frac{x}{5} \qquad \frac{13}{x} = \frac{x}{16}$$

$$x^2 = 77 \qquad x^2 = 20 \qquad x^2 = (13)(16)$$

$$x = \sqrt{77} \qquad x = \sqrt{20} = \sqrt{4 \cdot 5} \qquad x = \sqrt{13 \cdot 4 \cdot 4}$$

$$= 2\sqrt{5} \qquad = 4\sqrt{13}$$

2. $13\sqrt{2}$. Apply the first part of the Corollary to Theorem 1, which states that the length of the altitude is the geometric mean between the segments of the hypotenuse. Thus,

$$\frac{13}{q} = \frac{q}{26}$$

$$q^2 = (26)(13)$$

$$q = \sqrt{13 \cdot 13 \cdot 2} = 13\sqrt{2}$$

3. $13\sqrt{6}$. Apply the Pythagorean Theorem and use the value for q that you calculated in Prob. 2.

$$(13\sqrt{2})^2 + (26)^2 = y^2$$

$$338 + 676 = y^2$$

$$1014 = y^2$$

$$\sqrt{2 \cdot 3 \cdot 13 \cdot 13} = y$$

$$13\sqrt{6} = y$$

4. $13\sqrt{3}$. Apply the Pythagorean Theorem and use the value that you calculated for q in Prob. 2.

$$x^2 = q^2 + (13)^2$$

$$= (13\sqrt{2})^2 + (13)^2$$

$$= 338 + 169$$

$$= 507$$

$$= \sqrt{507}$$

$$= 13\sqrt{3}$$

5. 20. The diagonal of a rectangle with sides of 12 and 16 creates two congruent right triangles with the diagonal as hypotenuse, because all four angles of a rectangle are right (90°) angles. Therefore, you can apply the Pythagorean Theorem, using the side lengths of the rectangle as the lengths of the legs of the triangle:

$$(12)^2 + (16)^2 = (\text{length of the diagonal } D)^2$$
$$144 + 256 = D^2$$
$$\sqrt{400} = D$$
$$20 = D$$

6. $\sqrt{290}$. You have been given the lengths of two legs of a right triangle, so apply the Pythagorean Theorem:

$$(11)^2 + (13)^2 = h^2$$
$$\sqrt{121 + 169} = h$$
$$\sqrt{290} = h$$

The number 290 contains no perfect squares, aside from 1.

7. (*a*) 28; (*b*) $14\sqrt{3}$. A right triangle with one angle of 30° also has a remaining angle of 60°. Thus, this is a 30°-60°-90° triangle and you can apply Theorem 6 to determine the lengths of the leg and of the hypotenuse.

(*a*) The leg opposite the 30° angle is the shorter leg and, according to the theorem, the measure of the hypotenuse in this type of triangle is twice the measure of the shorter leg. So, the length of the hypotenuse equals twice times 14:

$$h = 2(14) = 28$$

(*b*) According to the theorem, the measure of the longer leg equals $\sqrt{3}$ times the length of the shorter leg. So, the length of the longer leg in this triangle is $\sqrt{3}$ times 14:

$$l = 14\sqrt{3}$$

8. $25\sqrt{2}$, 25. A right triangle with one angle measuring 45° also contains another angle measuring 45°, thus forming a 45°-45°-90° triangle. You can use the Pythagorean Theorem to calculate the measure of the hypotenuse or you can apply Theorem 5, which states that the measure of the hypotenuse in a 45°-45°-90° triangle is $\sqrt{2}$ times the measure of a leg.

Applying the Pythagorean Theorem:

$$(25)^2 + (25)^2 = h^2$$
$$2(25)^2 = h^2$$
$$\sqrt{2(25)^2} = h^2$$
$$25\sqrt{2} = h^2$$

Applying Theorem 5 takes less time, as long as you recall that the hypotenuse in a 45°-45°-90° triangle is $\sqrt{2}$ times the measure of the leg. The leg has been identified as 25, so the hypotenuse is equal to $25\sqrt{2}$.

9. (*a*) Draw a diagonal; (*b*) 160.

(*a*) To divide the square into two equal parts and to provide the longest possible running track, draw a diagonal. The result will be two isosceles triangles with side lengths of $85\sqrt{2}$.

(*b*) The diagonal will produce the longest possible track and you can calculate the length by applying Theorem 5, which states that the hypotenuse of an isosceles (45°-45°-90°) triangle is equal to $\sqrt{2}$ times the side of the triangle:

$$h = (85\sqrt{2})(\sqrt{2}) = 85(2) = 160$$

10. $4\sqrt{3}$. The altitude of an equilateral triangle bisects the side it intersects, forming two equilateral right triangles. Apply the Pythagorean Theorem to solve for the altitude, which is the longer leg of the two new triangles formed by the bisection:

$$a^2 + 4^2 = 8^2$$

$$a^2 + 16 = 64^2$$

$$a^2 = 64 - 16 = 48$$

$$a = 4\sqrt{3}$$

Chapter 9

Circles

A *circle* is a set of points in a plane at a given distance from a given point in that plane, the *center,* and it is the source of numerous calculations in geometry. The shape lends itself to many real-world associations. From automobile wheels to Ferris wheels, the movement of circular objects provides an interesting source of speculation regarding size and distance problems. The addition of lines that form angles or function as *tangents,* lines in the plane of the circle that intersect with the circle in exactly one point, requires that you also call upon skills related to angles, lines in a plane, and triangles.

Definitions to Know

Central angle. An angle that has its vertex at the center of a circle and whose interior measure forms a **minor arc** of the circle while the remainder of the circle is defined as the **major arc,** as in Fig. 9-1.

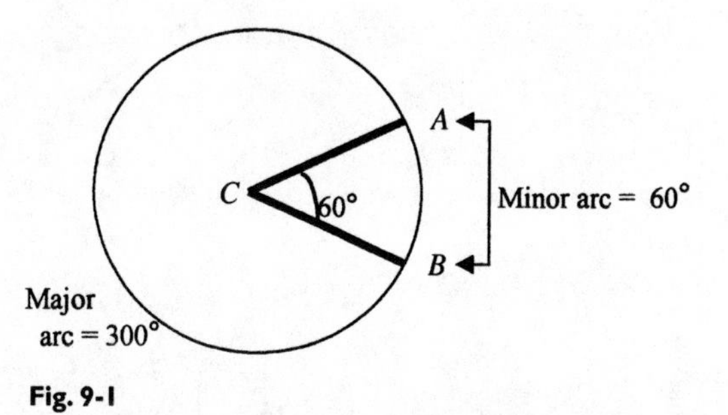

Fig. 9-1

Chord. A line segment with endpoints that lie on a circle, as in Fig. 9-2.

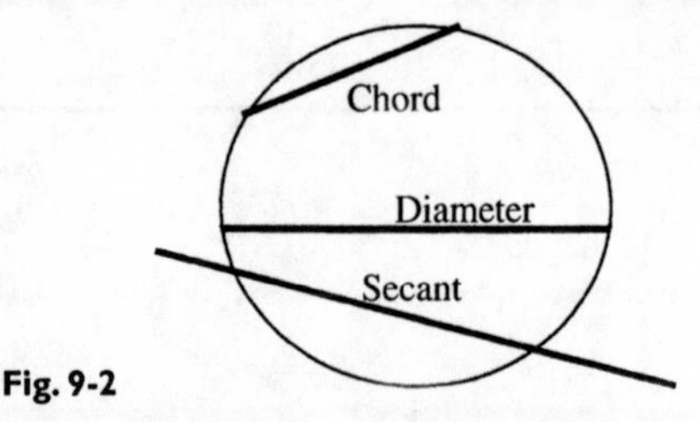

Fig. 9-2

Circumscribe. A circle is circumscribed around a polygon when all vertices of the polygon lie on the circle, as in Fig. 9-3.

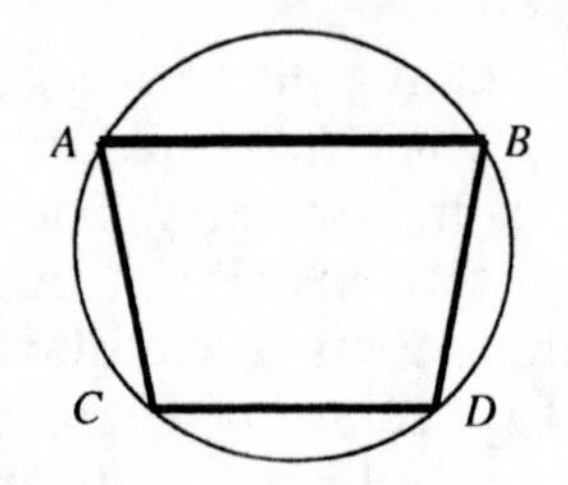

Fig. 9-3

Common tangent. A line that is tangent to two coplanar circles at different points and may be either internal, intersecting the segment that joins the centers, or external, not intersecting the segment that joins the centers, as in Fig. 9-4.

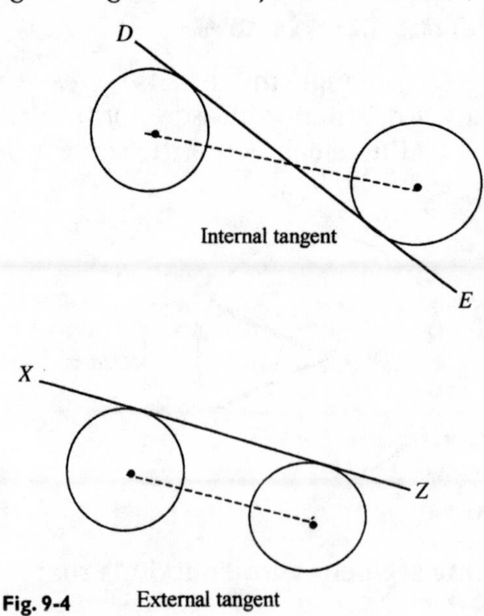

Fig. 9-4 External tangent

Concentric circles. Circles that lie in the same plane and have the same center, as in Fig. 9-5.

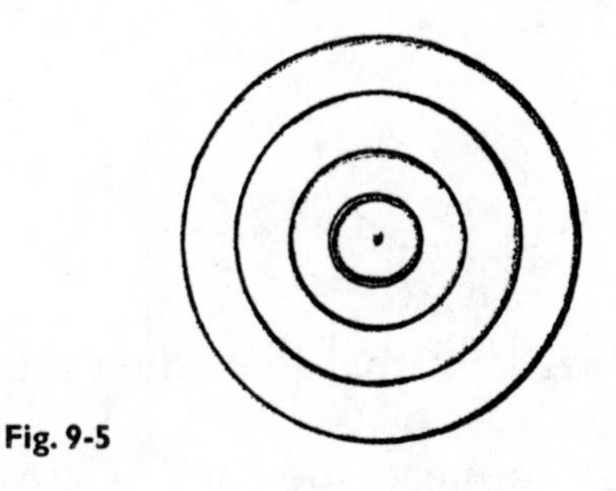
Fig. 9-5

Congruent arcs. Arcs that have equal measures.

Congruent circles. Circles that have congruent radii.

Diameter of the circle. A chord that contains the center of a circle, as in Fig. 9-2.

Inscribe. A polygon is inscribed in a circle when all vertices of the polygon lie on the circle, as in Fig. 9-3.

Inscribed angle. An angle with sides that are chords of the circle and whose vertex lies on the circle, as in Fig. 9-6.

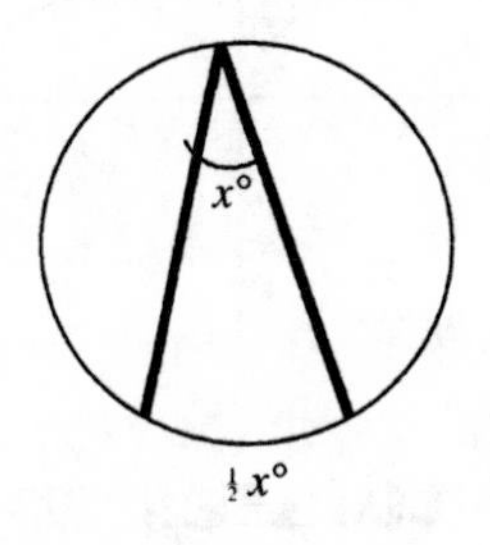

Fig. 9-6

Measure of a major arc. 360° minus the measure of the related minor arc, as in Fig. 9-1.

Measure of a minor arc. The same as the measure of its central angle, as in Fig. 9-1.

Point of tangency. The specific point where the circle and a tangent intersect, as in Fig. 9-7.

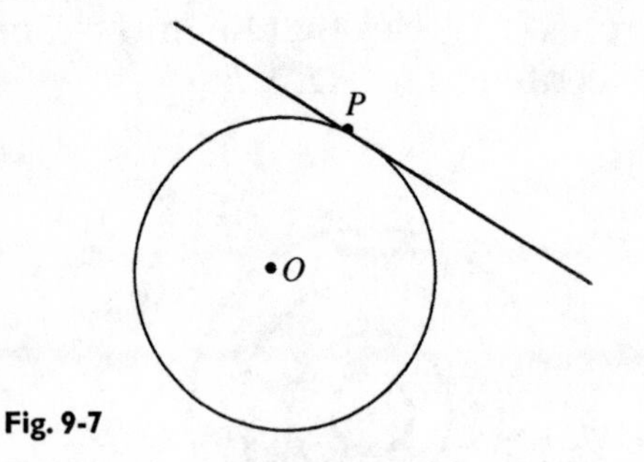

Fig. 9-7

Radius. Any line segment that joins the center of a circle to a point of the circle.

Secant. A line that connects the circle in two points and contains a chord, as in Fig. 9-2.

Semicircle. An arc with endpoints on a diameter of the circle, which measures 180°.

Tangent. The line in the plane of a circle that intersects the circle in exactly one point, as in Fig. 9-7.

Tangent circles. Coplanar circles that are tangent to the same line at the same point, as in Fig. 9-8.

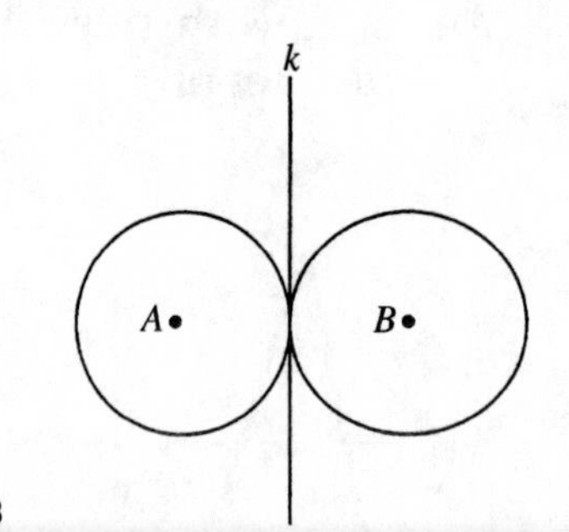

Fig. 9-8

Relevant Postulates and Theorems

Arc Addition Postulate

The measure of the arc formed by two adjacent arcs is the sum of the measures of these two arcs.

Theorem 1

If a line is tangent to a circle, then the line is perpendicular to the radius drawn to the point of tangency, as in Fig. 9-9.

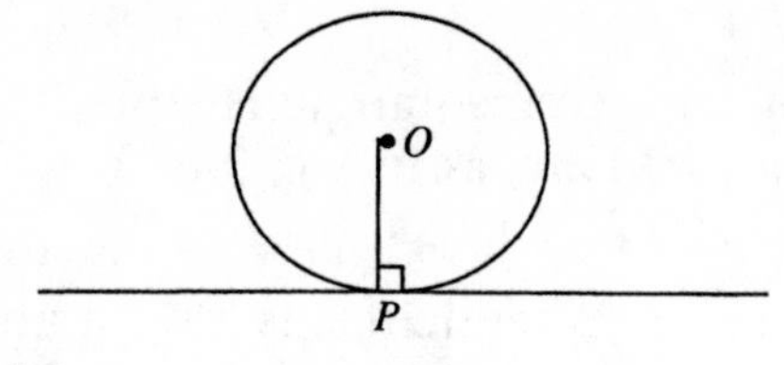

Fig. 9-9

Theorem 2

Tangents drawn to a circle from the same point are congruent, as in Fig. 9-10.

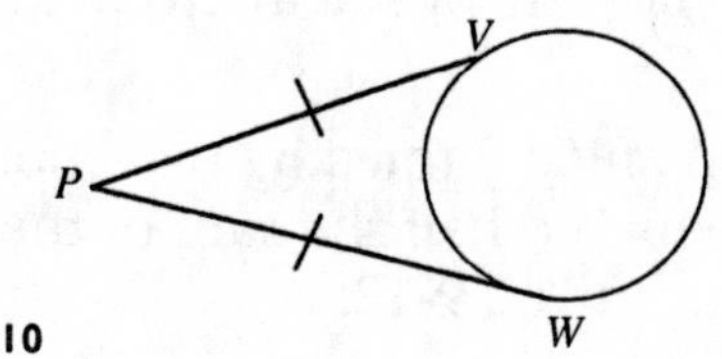

Fig. 9-10

Theorem 3

Two minor arcs are congruent if and only if their central angles are congruent.

Theorem 4

Congruent arcs have congruent chords, and congruent chords have congruent arcs.

Theorem 5

A diameter that is perpendicular to a chord bisects the chord and its arc, as in Fig. 9-11.

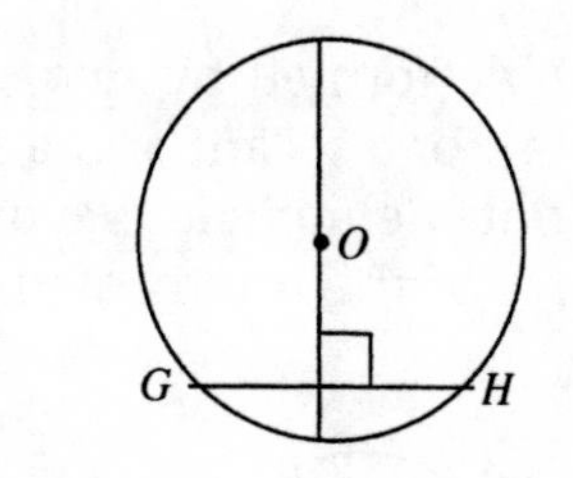

Fig. 9-11

Theorem 6

Chords that are equally distant from the center of a circle are congruent.

Theorem 7

The measure of an inscribed angle is equal to half the measure of its intercepted arc, as in Fig. 9-6.

Theorem 8

An angle inscribed in a semicircle is a right angle.

Theorem 9

If a quadrilateral is inscribed in a circle, then its opposite angles are supplementary.

Theorem 10

The measure of an angle formed by a chord and a tangent is equal to half the measure of the intercepted arc.

Theorem 11

The measure of an angle formed by two chords that intersect inside a circle is equal to half the sum of the measures of the intercepted arcs, as in Fig. 9-12.

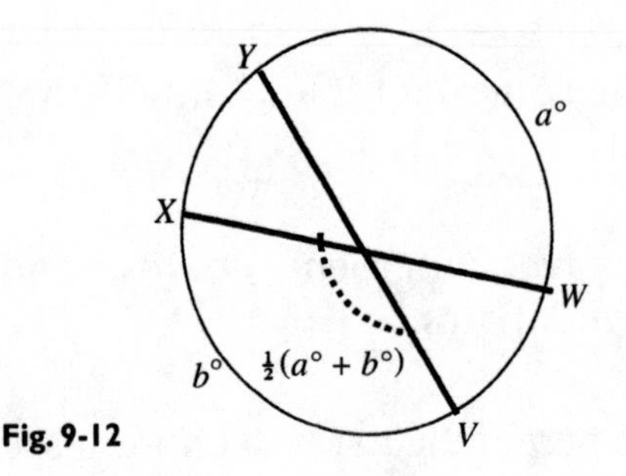

Fig. 9-12

Theorem 12

The measure of an angle formed by two secants (Fig. 9-13), two tangents (Fig. 9-14), or a secant and a tangent (Fig. 9-15) drawn from a point outside a circle is equal to half the difference of the measures of the intercepted arcs.

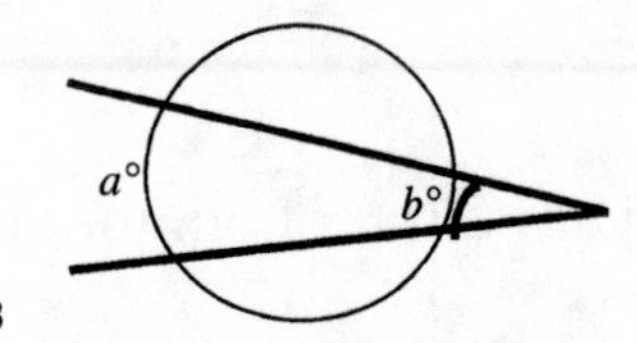

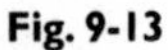

Fig. 9-13

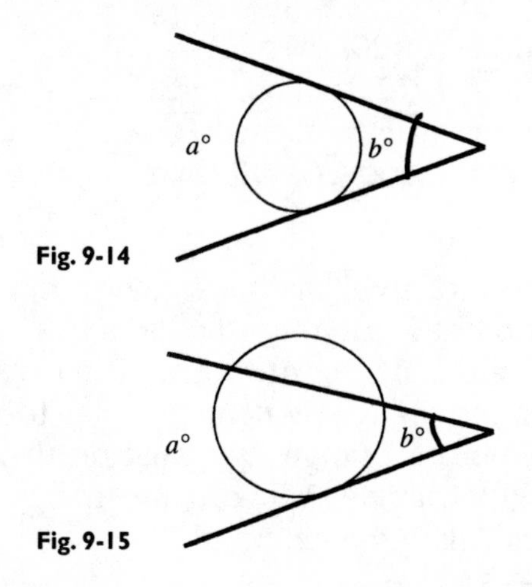

Fig. 9-14

Fig. 9-15

Theorem 13

When two chords intersect inside a circle, the product of the segments of one chord equals the product of the segments of the other chord.

Theorem 14

When two secant segments are drawn to a circle from an external point, the product of one secant segment and its external segment equals the product of the other secant segment and its external segment.

Theorem 15

When a secant segment and a tangent segment are drawn to a circle from an external point, the product of the secant segment and its external segment is equal to the square of the tangent segment.

EXAMPLE 1

In Fig. 9-16, if $\overline{FP}$ is tangent to circle *O* at point *P*, calculate the following measures in parts (*a*) through (*d*).

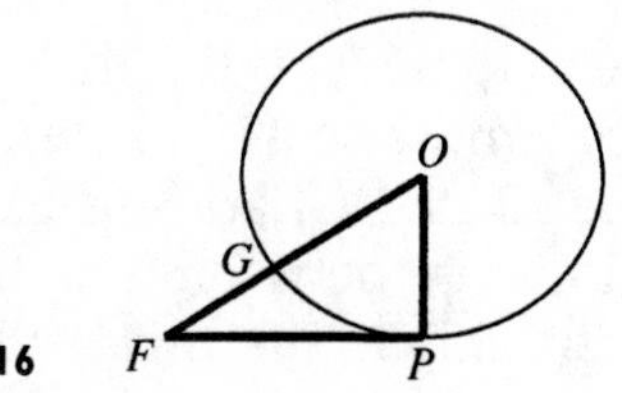

Fig. 9-16

(*a*) If $\overline{OP} = 12$ and $\overline{FO} = 20$, then $\overline{FP} =$ ______.
(*b*) If $\overline{OP} = 9$ and $\overline{FP} = 12$, then $\overline{FO} =$ ______.
(*c*) If $m\measuredangle POF = 60$ and $\overline{OP} = 9$, then $\overline{FO} =$ ______.
(*d*) If $\overline{FP} = 15$, $\overline{FG} = 8$, and $\overline{GO} = 9$ then $\overline{OP} =$ ______.

Solution

(*a*) The triangle formed by $\overline{FP}$, a line segment that is tangent to circle *O* at point *P*, intersects $\overline{OP}$, a radius drawn to *P*, the point of tangency. According to Theorem 1, a line that is tangent to a circle is perpendicular to the radius drawn to the point of tangency. Based on this theorem, $\triangle FOP$ is a right triangle. You can use the Pythagorean Theorem to calculate the length of $\overline{FP}$.

$$(\overline{FP})^2 + (\overline{OP})^2 = (\overline{FO})^2$$
$$(\overline{FP})^2 + (12)^2 = (20)^2$$
$$(\overline{FP})^2 + 144 = 400$$
$$(\overline{FP})^2 = 400 - 144 = 256$$
$$\overline{FP} = 256 = 16$$

(*b*) The triangle formed by $\overline{FP}$, a line segment that is tangent to circle *O* at point *P*, intersects $\overline{OP}$, a radius drawn to *P*, the point of tangency. According to Theorem 1, a line that is tangent to a circle is perpendicular to the radius drawn to the point of tangency. Based on this theorem, $\triangle FOP$ is a right triangle. You can use the Pythagorean Theorem to calculate the length of $\overline{FO}$.

$$(\overline{FP})^2 + (\overline{OP})^2 = (\overline{FO})^2$$
$$(12)^2 + (9)^2 = (\overline{FO})^2$$
$$144 + 81 = (\overline{FO})^2$$
$$225 = (\overline{FO})^2$$
$$15 = \overline{FO}$$

(*c*) Applying Theorem 1, $\triangle FOP$ is a right triangle. If $m\measuredangle POF$ is 60°, then $m\measuredangle OFP$ must be 30°, making $\triangle FOP$ a 30°-60°-90° right triangle. Recall what you learned in Chap. 8 about the relationship between the legs and the hypotenuse. In a 30°-60°-90° triangle, the hypotenuse is twice the length of the shortest leg, which in this case is $\overline{OP}$. The longest leg is $\sqrt{3}$ times the length of the shortest leg, which in this case also $\overline{OP}$.

In $\triangle FOP$, $\overline{FO}$ is the hypotenuse and $\overline{OP}$ is the shortest side. Thus,

$$\overline{FO} = 2(\overline{OP}) = 2(9) = 18$$

(*d*) Use the Pythagorean Theorem to calculate the measure of leg $\overline{OP}$. First, find the measure of the hypotenuse = $\overline{FO}$.

$$\overline{FG} + \overline{GO} = \overline{FO}$$
$$8 + 9 = 17 = \overline{FO}$$

Then, applying the Pythagorean Theorem, the following is true:

$$(\overline{FP})^2 + (\overline{OP})^2 = (\overline{FO})^2$$
$$(15)^2 + (\overline{OP})^2 = (17)^2$$
$$(\overline{OP})^2 = (17)^2 - (15)^2 = 289 - 225 = 64$$
$$\overline{OP} = 8$$

EXAMPLE 2

In Fig. 9-17, find the measure of the following arcs:

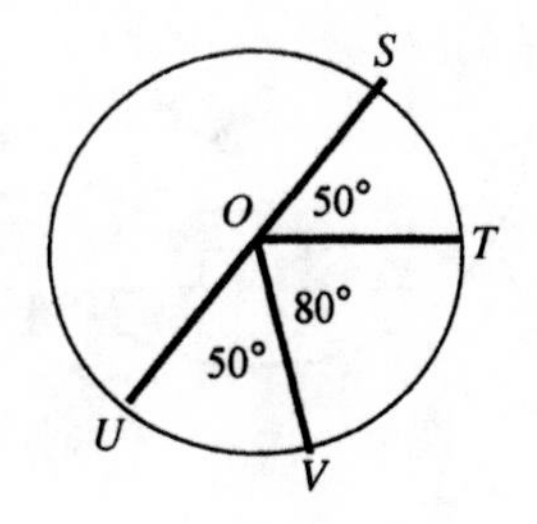

Fig. 9-17

(*a*) $\widehat{UV}$
(*b*) $\widehat{UT}$
(*c*) $\widehat{UVTS}$
(*d*) $\widehat{STV}$

Solution

(*a*) $\overset{\frown}{UV}$ is a minor arc and, by definition, the measure of a minor arc is equal to the measure of its central angle. In Fig. 9-17, the central angle of $\overset{\frown}{UV}$ is $\measuredangle UOV$, and $m\measuredangle UOV = 50°$. Thus, the measure of $\overset{\frown}{UV} = 50°$.

(*b*) $\overset{\frown}{UT}$ is formed by two adjacent arcs. Applying the Arc Addition Postulate, the measure of an arc formed by two adjacent arcs is equal to the sum of the measures of these two arcs.

$$\overset{\frown}{UT} = \overset{\frown}{UV} = \overset{\frown}{VT}$$

$$\text{Measure of } \overset{\frown}{UT} = 50° + 80° = 130°$$

(*c*) $$\overset{\frown}{UVTS} = \overset{\frown}{UV} + \overset{\frown}{VT} + \overset{\frown}{TS} \text{ (a semicircle)}$$

Measure of $\overset{\frown}{UVTS} = 50° + 80° + 50° = 180°$ (a semicircle)

(*d*) $$\overset{\frown}{STV} = \overset{\frown}{ST} + \overset{\frown}{TV}$$

$$\text{Measure of } \overset{\frown}{STV} = 50° + 80° = 130°$$

EXAMPLE 3

In Fig. 9-18, *O* is the center of the circle, $\overline{AD}$ is the diameter and chords *AB* and *CD* are perpendicular to chord *AC* and parallel to each other. Find the values below.

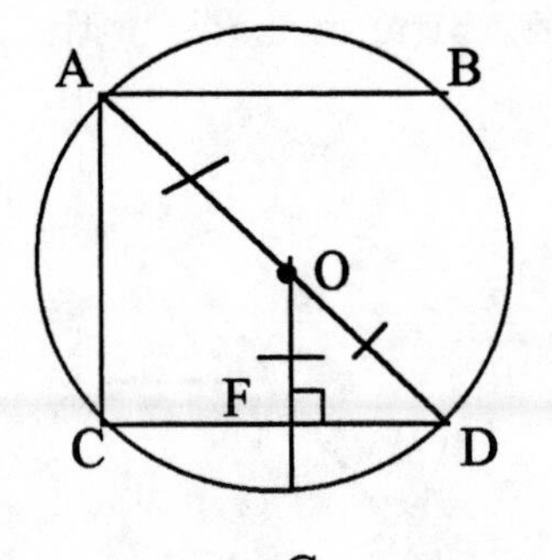

Fig. 9-18

(*a*) If $\measuredangle DOG = 45°$, then the measure of $\overset{\frown}{GD} =$ ______ and the measure of $\overset{\frown}{DBG} =$ ______.

(*b*) If $\overline{OF} = 5$ and $\overline{OA} = 8$, what is the length of $\overline{CD}$?

(*c*) If $\overline{AD}$ is the diameter of a circle with center *O*, what is the measure of $\measuredangle ACD$?

(*d*) If $m\measuredangle ADC = 45°$, what is $m\measuredangle BAD$?
(*e*) What is the measure of $\overset{\frown}{BD}$?

Solution

(*a*) $\measuredangle DOG$ is a central angle and $\overset{\frown}{GD}$ is a minor arc. By definition, the measure of a minor arc is equal to the measure of its central angle. Thus, the measure of $\overset{\frown}{GD}$ is 45°.

The measure of the major arc of a central angle is 360° minus the measure of the minor arc. $\overset{\frown}{DBG}$ is a major arc of $\measuredangle DOG$, so its measure is $360° - 45° = 315°$.

(*b*) $\overline{AO} \cong \overline{OD} = 8$ is given, and $\triangle OFD$ is a right triangle. According to Theorem 5, a radius perpendicular to a chord bisects the chord and its arc. Therefore, the measure of $\overline{FD} = \frac{1}{2}\overline{CD}$. Apply the Pythagorean Theorem to solve for the measure of $\overline{FD}$. Thus,

$$(\overline{OF})^2 + (\overline{FD})^2 = (\overline{OD})^2$$

$$5^2 + (\overline{FD})^2 = 8^2$$

$$(\overline{FD})^2 = 64 - 25$$

$$\overline{FD} = \sqrt{39}$$

But, $$\overline{FD} = \tfrac{1}{2}\overline{CD}$$

So, $$\overline{CD} = 2(\overline{FD}) = 2\sqrt{39}$$

(*c*) $\measuredangle ACD$ is inscribed in a semicircle. According to Theorem 8, a triangle inscribed in a semicircle is a right triangle. Thus, the measure of $\measuredangle ACD = 90°$.

(*d*) We are given the information that $\overline{AB}$ is parallel to $\overline{CD}$. Therefore, they form parallel lines cut by the transversal of the diameter $\overline{AD}$. Thus, $\measuredangle BAD$ and $\measuredangle ABC$ are alternate interior angles and, by definition, they are equal in measure. So, $m\measuredangle BAD = 45°$.

(*e*) $\measuredangle BAD$ is an inscribed angle in circle *O*. According to Theorem 7, the measure of an inscribed angle is equal to one-half the measure of the intercepted arc. Using the definition of parallel lines cut by a transversal, we have already determined that $m\measuredangle BAD$, which intercepts this arc, is 45°. Thus the measure of the arc is twice that measure, or 90°.

Supplementary Circles Problems

1. Complete the table using Fig. 9-19.

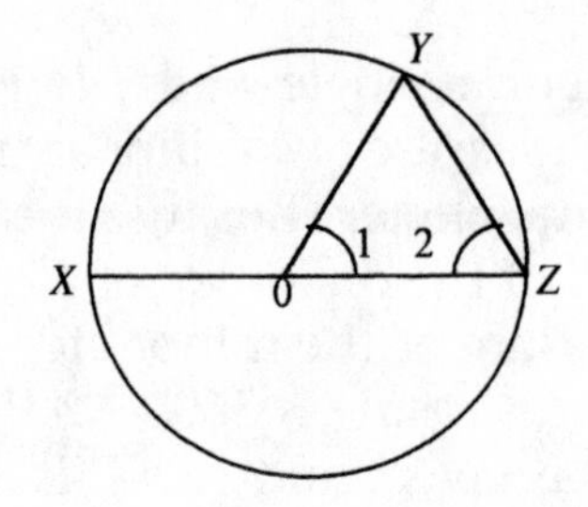

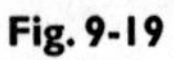

Fig. 9-19

$m\widehat{YZ}$	40°	50°			x°
$m\measuredangle YOZ$			58°		
$m\measuredangle OZY$				25	

2. In Fig. 9-20, $\measuredangle RQS$ is a right angle inscribed in semicircle RS.

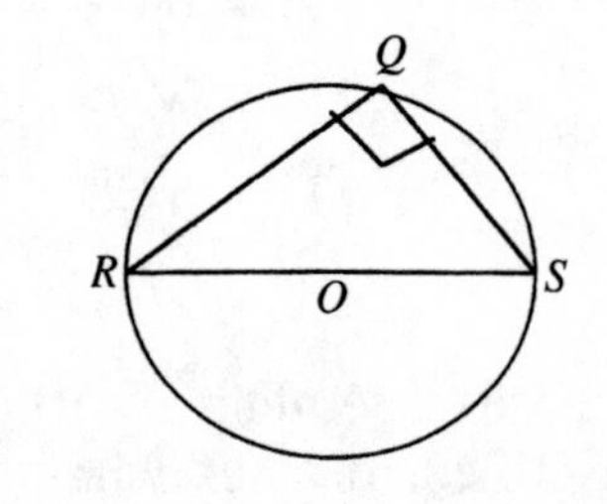

Fig. 9-20

(*a*) If the measure of $\widehat{RQ}$ is equal to the measure of $\widehat{QS}$, what is the measure of $\measuredangle QSR$?

(*b*) If $\overline{RS} = 10$, what is the measure of $\overline{QS}$?

3. In Fig. 9-21, what is the measure of $\measuredangle x$?

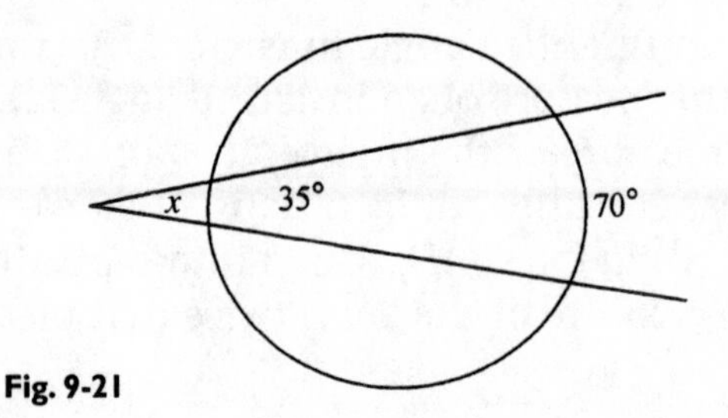

Fig. 9-21

4. In Fig. 9-22, what is the value of z?

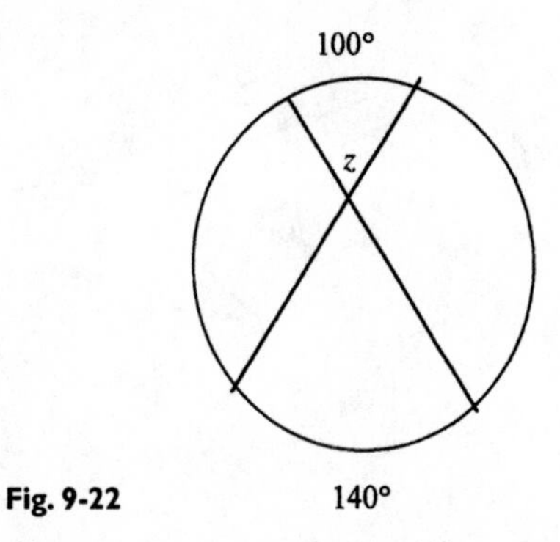

Fig. 9-22

5. In Fig. 9-23, if the measure of $\widehat{ACB} = 230°$, then the measure of $\measuredangle P$ is ______.

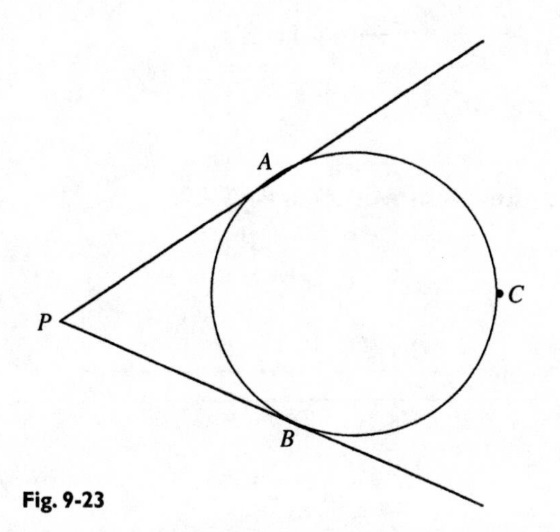

Fig. 9-23

6. In Fig. 9-24, $m \measuredangle x = 70°$, and $c = 40°$, then $b =$ ______.

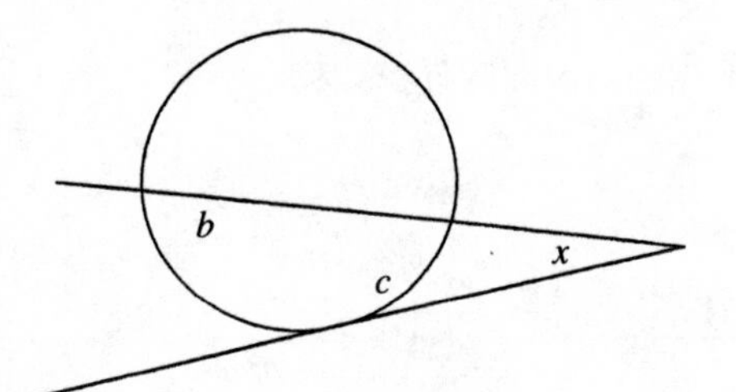

Fig. 9-24

7. Solve for y in Fig. 9-25.

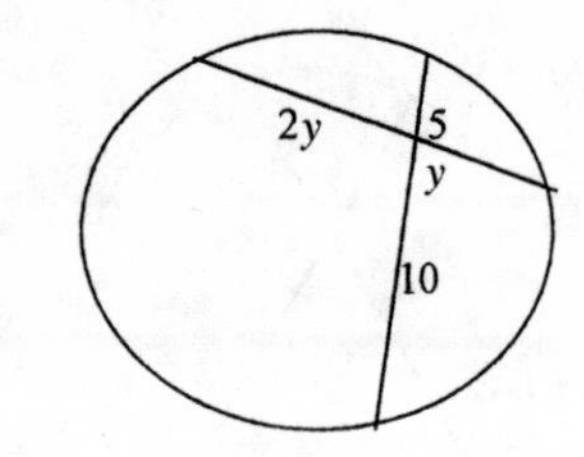

Fig. 9-25

8. In Fig. 9-26, solve for x.

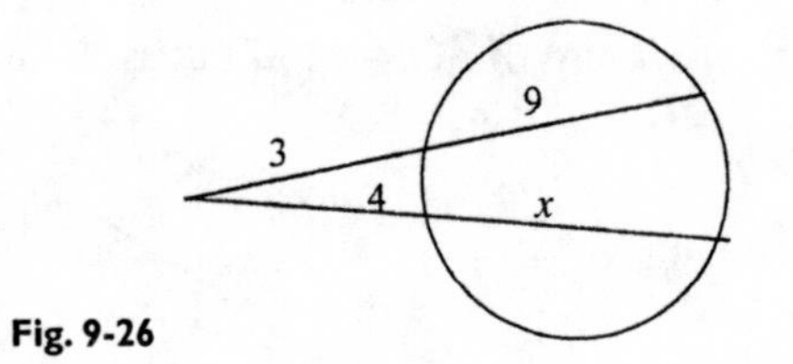

Fig. 9-26

9. What is the measure of $\measuredangle 1$ in Fig. 9-27?

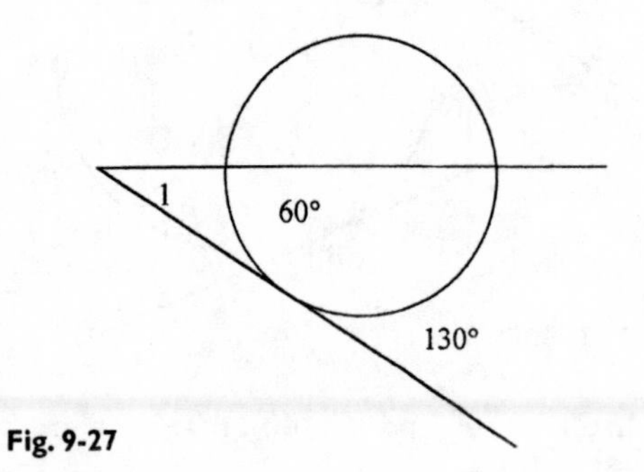

Fig. 9-27

10. Identify the value of b in Fig. 9-28.

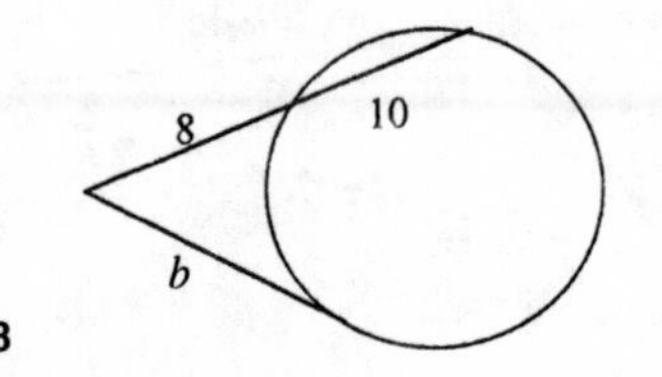

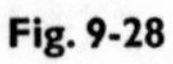

Fig. 9-28

Solutions to Supplementary Circles Problems

1.

$m\widehat{YZ}$	40°	50°	58°	65	$x°$
$m\measuredangle 1$	40°	50°	58°	65	$x°$
$m\measuredangle 2$	50°	40°	32°	25	$90 - x°$

In Fig. 9-19, $\measuredangle 1$ is a central angle. Based on definition, the measure of a minor arc is equal to the measure of its central angle. Therefore, the measures of the minor arc $\widehat{YZ}$ are equal in degrees to the measures of the central angle $\measuredangle 1$ in the above table.

In Fig. 9-19, $\measuredangle OYZ$ is an angle inscribed in a semicircle. According to Theorem 8, an angle inscribed in a semicircle is a right angle with the vertex of the right angle on the circle. Thus, $m\measuredangle 1 + m\measuredangle 2 = 90°$.

2. (*a*) If the two minor arcs are equal, then the measures of their angles are equal. Thus, because we know that $\measuredangle RQS$ is a right (90°) angle, we know that the sum of the measures of $\measuredangle QRS$ and $\measuredangle QSR$ must be 90°. Therefore, $m\measuredangle QRS = m\measuredangle QSR = 45°$.

(*b*) We are given the fact that $\triangle RQS$ is a right triangle, and we know that it is also an isosceles right triangle, so we can use the Pythagorean Theorem to find the measure of the legs of the triangle. We already know that the measure of the hypotenuse (also the diameter $\overline{RS}$) is 10. Thus,

$$a^2 + a^2 = (10)^2$$
$$2a^2 = 100$$
$$a^2 = 50$$
$$a = \sqrt{50} = \sqrt{25 \cdot 2} = 5\sqrt{2}$$

3. In Fig. 9-21, $\measuredangle x$ is formed by two secants. According to Theorem 12, the measure of an angle formed by two secants is equal to half the differences of the measures of the intercepted arcs. Thus, $m\measuredangle x = \frac{1}{2}(70° - 35°) = 17\frac{1}{2}°$.

4. In Fig. 9-22, $\measuredangle z$ is formed by the intersection of two chords that intersect inside a circle. According to Theorem 13, the measure of $\measuredangle z$ is equal to half the sum of the intercepted arcs. Thus,

$$m\measuredangle z = \tfrac{1}{2}(100 + 140) = \tfrac{1}{2}(240) = 120°$$

5. In Fig. 9-23, if the measure of $\overset{\frown}{ACB} = 230°$, then the measure of $\overset{\frown}{AB} = 130°$. $\measuredangle P$ is formed by two tangents, so use Theorem 12 to find the measure of $\measuredangle P$. According to Theorem 12, the measure of an angle formed by two tangents is equal to half the difference of the measures of the intercepted arcs. Thus,

$$m\measuredangle P = \tfrac{1}{2}(230 - 130) = \tfrac{1}{2}(100) = 50°$$

6. In Fig. 9-24, $\measuredangle x$ is formed by a tangent and a secant drawn from a point outside the circle. According to Theorem 12, the measure of the angle is equal to half the difference of the measures of the intercepted arcs. Thus,

$$\begin{aligned} m\measuredangle x &= \tfrac{1}{2}(b° - c°) \\ 70 &= \tfrac{1}{2}(b - 40) = \tfrac{1}{2}b - 20 \\ 90 &= \tfrac{1}{2}b \\ 180° &= b \end{aligned}$$

7. Figure 9-25 contains the intersection of two chords inside a circle. To solve for y, apply Theorem 13, which states that the product of the segments of one chord equals the product of the segments of the other chord. Thus,

$$\begin{aligned} 10 \cdot 5 &= 2y \cdot y \\ 50 &= 2y^2 \\ 25 &= y^2 \\ 5 &= y \end{aligned}$$

8. In Fig. 9-26, two secant segments are drawn to a circle from an external point. To solve for x, apply Theorem 14, which states that the product of one secant segment and its external segment equals the product of the other secant segment and its external segment. Thus,

$$\begin{aligned} 3\,(3 + 9) &= 4\,(4 + x) \\ 3\,(12) &= 16 + 4x \\ 36 - 16 &= 4x \\ 20 &= 4x \\ 5 &= x \end{aligned}$$

9. In Fig. 9-27, $\measuredangle 1$ is formed by a secant and a tangent drawn from a point outside the circle. According to Theorem 12, the measure of

the angle is equal to half the difference of the measures of the intercepted arcs. Thus,

$$m\measuredangle 1 = \tfrac{1}{2}(130° - 60°) = \tfrac{1}{2}(70°) = 35°$$

10. In Fig. 9-28, a secant segment and a tangent segment are drawn to a circle from an external point. To solve for b, apply Theorem 15, which states that the product of a secant segment and its external segment is equal to the square of the tangent segment. Thus,

$$\begin{aligned} 8(8 + 10) &= b^2 \\ 8(18) &= b^2 \\ 144 &= b^2 \\ 12 &= b \end{aligned}$$

Chapter 10

Areas of Polygons and Circles

The formulas for determining the areas of rectangles, parallelograms, triangles, rhombuses, trapezoids, and regular polygons require that you remember and apply the skills that we reviewed in Chap. 1. In the same way that the lengths of line segments and the measures of angles are expressed as positive numbers, the *area of a polygon,* the region that includes the polygon and its interior, is also expressed in positive numbers. Many of the same skills used in determining measure in earlier chapters must be applied in calculating area. One important difference is that the many types of polygons require you to memorize formulas for each, but we will forgo the proofs and, instead, move on to application.

Definitions to Know

Altitude to a base. Any segment perpendicular to the line segment representing the base drawn from any point on the opposite side. All altitudes to the same base have the same length.

Apothem of a regular polygon. The perpendicular distance from the center of the polygon to a side, such as segment $\overline{OY}$ in Fig. 10-1.

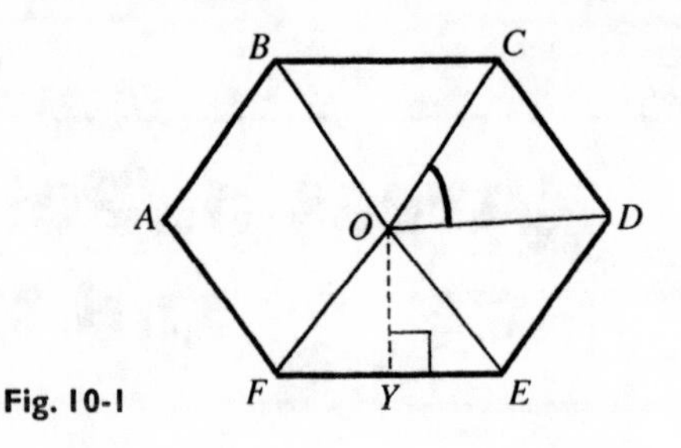

Fig. 10-1

Base. Term used to refer to the line segment or the length of any line segment that functions as any side of a rectangle or other parallelogram.

Central angle of a regular polygon. An angle formed by two radii drawn to consecutive vertices, such as $\measuredangle COD$ in Fig. 10-1.

Height of a polygon. The length of an altitude. All the altitudes to a particular base have the same length.

Measure of a central angle of a regular polygon. Equal to 360° divided by the number of sides.

Perimeter of a polygon. Equal to the sum of the lengths of its sides.

Radius of a regular polygon. The distance from the center of the polygon to any vertex, such as $\overline{OB}$, $\overline{OC}$, $\overline{OD}$, and $\overline{OE}$ in Fig. 10-1.

Regular polygon. A polygon that is both equiangular and equilateral, around which you can circumscribe a circle.

Relevant Postulates and Theorems

Postulate 1

The area of a square is the product of a side and itself ($A = s^2$).

Postulate 2

Two figures that are congruent have the same area.

Postulate 3

The area of a region is the sum of the areas of its nonoverlapping parts, as in Fig. 10-2, in which the area of figure $WXYZ$ is equal to area 1 + area 2 + area 3.

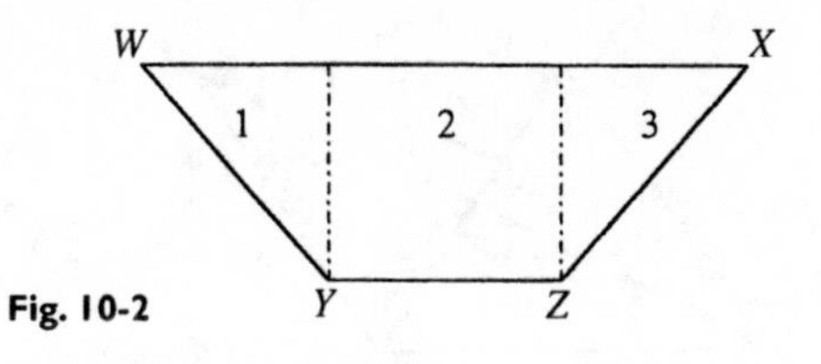

Fig. 10-2

Theorem 1

The area of a rectangle is equal to the product of the base and the height ($A = bh$).

Theorem 2

The area of a parallelogram is equal to the product of a base and the length of an altitude to that base (the height), as in Fig. 10-3 ($A = bh$).

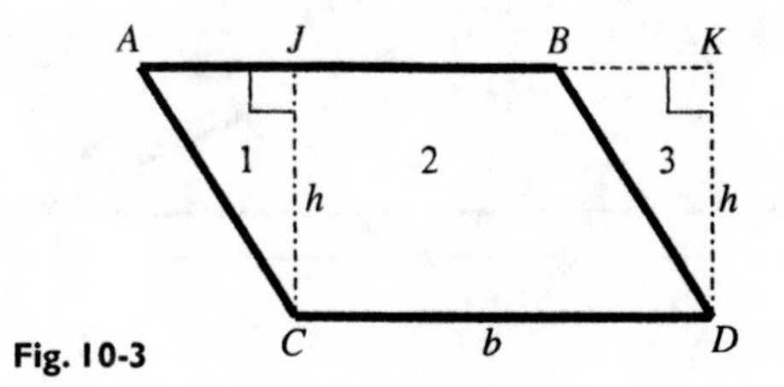

Fig. 10-3

Theorem 3

The area of a triangle is equal to one-half the product of a base and the length of an altitude to that base (the height), as in Fig. 10-4 ($A = \frac{1}{2}bh$).

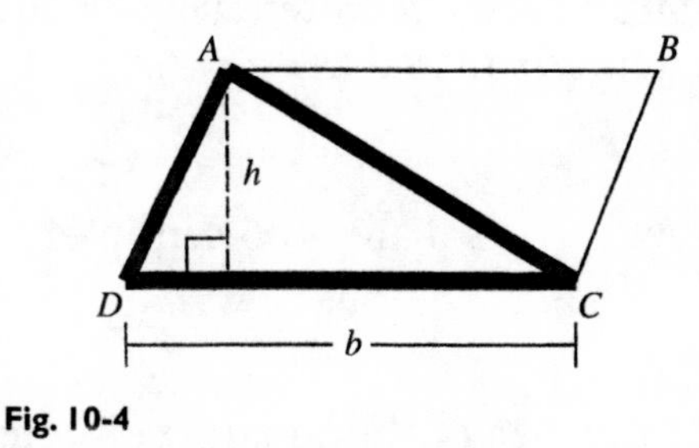

Fig. 10-4

Theorem 4

The area of a rhombus is equal to one-half the product of the length of its diagonals, as in Fig. 10-5 ($A = \frac{1}{2}d_1d_2$).

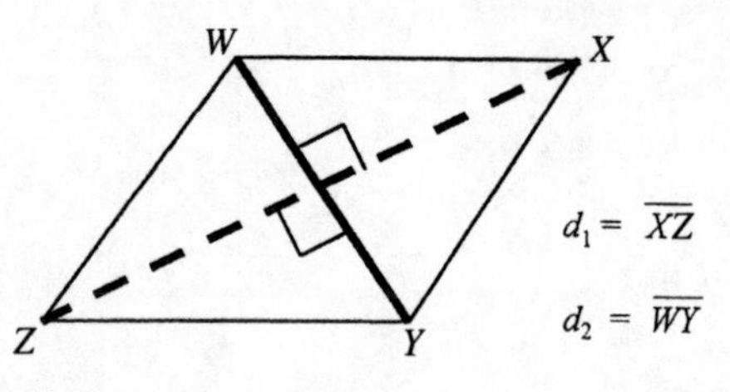

Fig. 10-5

Theorem 5

The area of a trapezoid equals one-half the product of the height and the sum of the bases, as in Fig. 10-6 [$A = \frac{1}{2}h(b_1 + b_2)$].

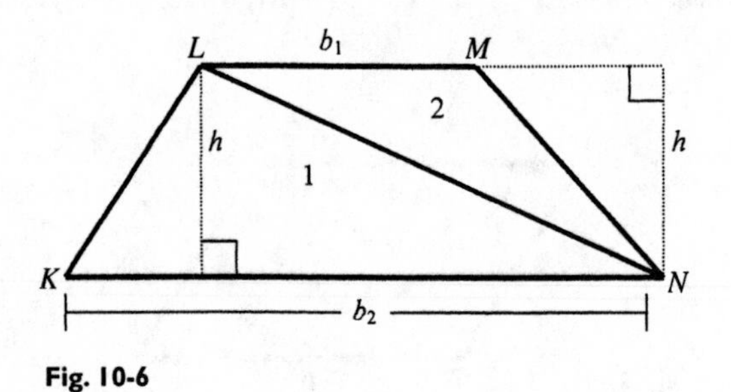

Fig. 10-6

Theorem 6

The area of a regular polygon is equal to one-half the product of the apothem and the perimeter ($A = \frac{1}{2}aP$).

Theorem 7

The area of a circle is equal to the product of the approximation for π, the ratio of the circumference of a circle to the diameter, and the radius squared ($A = \pi r^2$).

EXAMPLE 1

If the perimeter of a square is 36 centimeters (cm), what is the area?

Solution

To calculate the area of the square, you must first determine the measure of a side of the square. The sides of a square are of equal length, and the perimeter of a square is equal to the sum of the four sides. Therefore, the length of each side s is equal to $36/4 = 9$ cm.

The area of a square is equal to the square of a side. Thus,

$$A = (9)^2 \text{ cm}^2 = 81 \text{ cm}^2$$

EXAMPLE 2

Complete the following table, calculating all measures for rectangles.

b	5 cm		$3\sqrt{5}$ cm		$(x + 3)$ cm
h	4 cm	7 cm	6 cm	$5\sqrt{7}$ cm	x cm
A		21 cm^2		70 cm^2	

Solution

To calculate the base (b), the height (h), and the area (A) measures in the above table, apply Theorem 1 and use the formula for the area of a rectangle: $A = bh$.

b	5 cm	3 cm	$3\sqrt{5}$ cm	$2\sqrt{7}$ cm	$(x + 3)$ cm
h	4 cm	7 cm	6 cm	$5\sqrt{7}$ cm	x cm
A	20 cm^2	21 cm^2	$18\sqrt{5}$ cm^2	70 cm^2	$(x^2 + 3x)$ cm^2

EXAMPLE 3

Find the area of $\triangle JKL$, which has side lengths of 5, 5, and 6, as in Fig. 10-7.

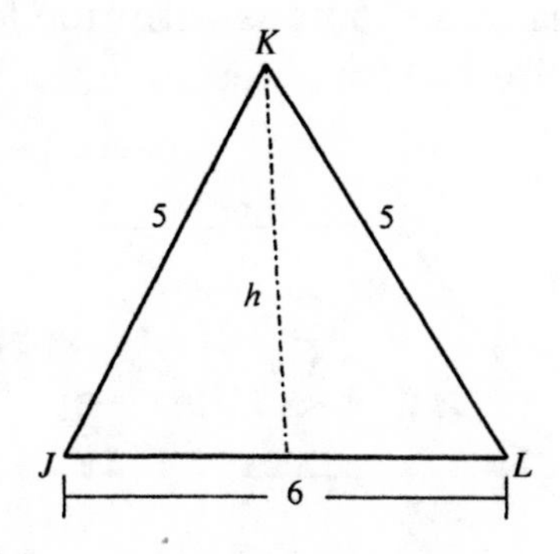

Fig. 10-7

Solution

A triangle with side lengths of 5, 5, and 6 is an isosceles triangle. Before you can calculate the area of the triangle, you have to determine the height, which you can do by drawing a perpendicular bisector from the vertex opposite the base to the base. Thus, you divide the triangle into two congruent triangles, each having a hypotenuse of 5 and a base of 3. Their shared side is the height (h) of the original triangle. Use the Pythagorean Theorem to solve for h.

$$3^2 + h^2 = 5^2$$
$$h^2 = 25 - 9$$
$$h^2 = 16$$
$$h = 4$$

To calculate the area of the triangle, apply Theorem 3 and substitute the values in the formula:

$$\tfrac{1}{2}bh = A$$

Keep in mind that you must use the measure of the base of the original triangle with the newly calculated height. Thus,

$$\tfrac{1}{2}(6)(4) = A$$
$$\tfrac{1}{2}(24) = A$$
$$12 = A$$

EXAMPLE 4

Find the area of a rhombus with side lengths of 17 and one diagonal that measures 30, as in Fig. 10-8.

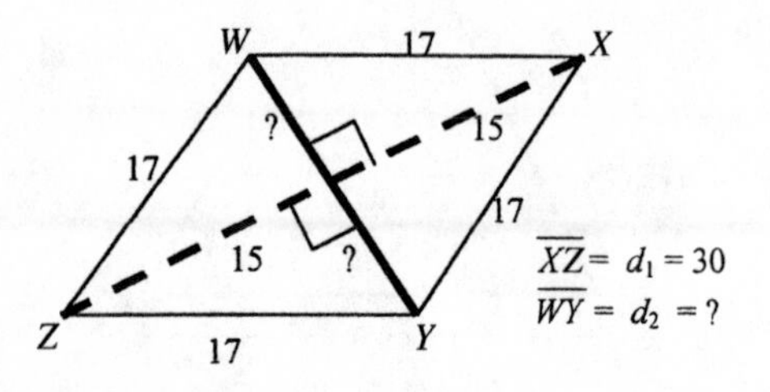

Fig. 10-8

Solution

We know from previous chapters that the sides of a rhombus are congruent and the diagonals of the rhombus bisect each other and form right (90°) angles. Using this information, we can calculate the measure of the second diagonal. Rhombus *WXYZ* contains four congruent right triangles. We can use the Pythagorean Theorem to solve for their bases to determine the measure of d_2.

$$(15)^2 + x^2 = (17)^2 \qquad \text{(where } x \text{ equals one-half the measure of } d_2\text{)}$$

$$225 + x^2 = 289$$

$$x^2 = 289 - 225 = 64$$

$$x = 8$$

Therefore, because x is equal to one-half the measure of d_2, then $d_2 = 16$.

To calculate the area of the rhombus, we can now apply Theorem 4, which states that the area of a rhombus is equal to one-half the product of the diagonals. Thus,

$$A = \tfrac{1}{2} d_1 d_2$$

$$= \tfrac{1}{2}(16)\,(30)$$

$$= \tfrac{1}{2}(480) = 240 \text{ square units}$$

EXAMPLE 5

Find the area of a trapezoid having bases of lengths 4 and 10 and legs of length 7, as in Fig. 10-9.

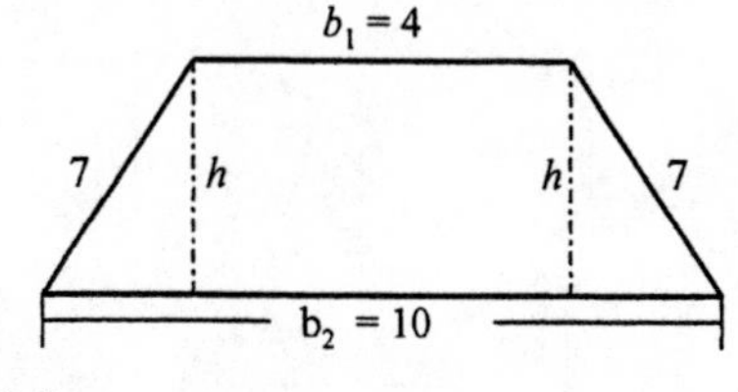

Fig. 10-9

Solution

To calculate the area of the trapezoid, apply Theorem 5, which states that the area of a trapezoid is equal to one-half the product of the height and sum of the bases. Before you can apply the formula, however, you must first find the height of the trapezoid.

In the trapezoid in Fig. 10-9, draw two altitudes perpendicular to b_2, as shown by the broken lines. The construction of the altitudes divides the figure into a rectangle and two congruent right triangles. The lower base has been divided into segments of lengths 3, 4, and 3, and we can use these figures and the Pythagorean Theorem to solve for h. Thus,

$$h^2 + 3^2 = 7^2$$
$$h^2 = 49 - 9 = 40$$
$$h = \sqrt{40} = \sqrt{4 \cdot 10} = 2\sqrt{10}$$

We can now apply the formula from Theorem 5 to find the area of a trapezoid:

$$A = \tfrac{1}{2}h(b_1 + b_2)$$
$$= \tfrac{1}{2}(2\sqrt{10})\,(4 + 10)$$
$$= \tfrac{1}{2}(2\sqrt{10})\,(14)$$
$$= 14(\sqrt{10})$$

EXAMPLE 6

Calculate the area of a regular octagon with sides of length 6 and radii of length 5, as in Fig. 10-10.

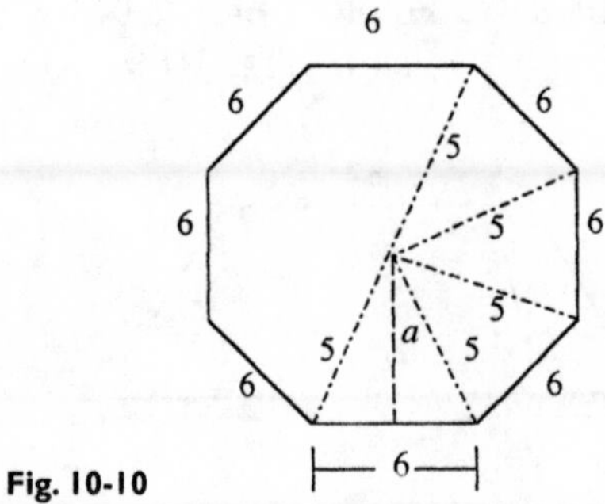

Fig. 10-10

Solution

To calculate the area of a regular octagon, apply Theorem 6, which states that the area is equal to one-half the product of the apothem (a) and the perimeter (P). Before you can apply the formula, you will have to calculate the measures of the apothem and the perimeter.

In Fig. 10-10, each triangular section of the octagon contains sides of 5 and a base of 6, which we will use to determine the measure of the apothem. When we draw an altitude from the center of the octagon (the vertex of one triangle), we create two congruent right triangles within each of the eight divisions. Each newly formed triangle has a base of 3 and a hypotenuse of 5. Use the Pythagorean Theorem to calculate the remaining leg, which is the apothem of the octagon, or simply recognize this as a 3-4-5 right triangle. Thus, the length of the apothem is equal to 4.

The perimeter (P) of the octagon is equal to the product of the measure of the sides and the number of the sides. Thus, $P = (6)\,(8) = 48$.

We now have enough information to calculate the area of the octagon using the formula provided in Theorem 6. Thus,

$$A = \tfrac{1}{2}aP = \tfrac{1}{2}(4)\,(48) = \tfrac{1}{2}(192) = 96$$

EXAMPLE 7

Find the area of a circle with a radius of 7 meters (m).

Solution

To calculate the area of the circle, apply Theorem 7, which states that the area of a circle is equal to the product of π and the square of the radius. The value for π is only approximate, and you may use either 22/7 or 3.1416 in your calculations. Thus,

$$A = \pi(r)^2 = \frac{22}{7}(7\text{ m})^2 = 22(7)(\text{m})^2 = 154\text{ m}^2$$

Supplementary Areas of Polygons and Circles Problems

1. Find the area of a square playing field if the length of the fence surrounding it is 320 feet (ft).
2. What is the area of a square with a diagonal measuring 25?

3. Find the length of the longer diagonal in a rhombus with sides of length 12 and a smaller diagonal of length 8.
4. What is the area of the rhombus in Prob. 3?
5. Find the area of a trapezoid that has bases measuring 11 feet and 23 feet and legs measuring 8. (See Fig. 10-11.)

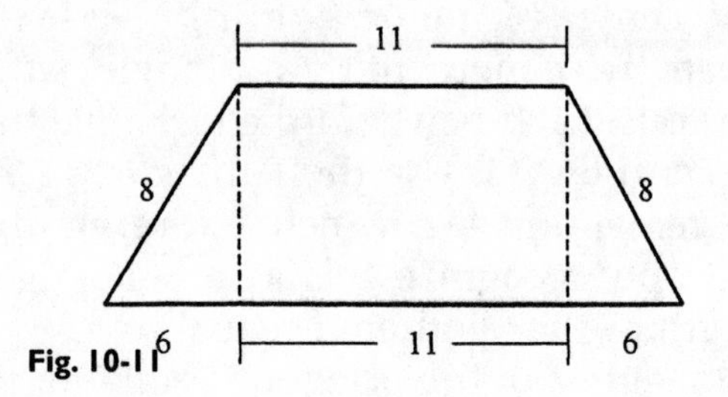

Fig. 10-11

For Probs. 6 through 9 complete the chart to show how the radius, circumference, and area of a circle change as different elements of the circle change. Leave the answers in terms of π.

6 through 9.

	Prob. 6	Prob. 7	Prob. 8	Prob. 9
Radius	21		$5\sqrt{3}$	
Circumference		18π		
Area				49π

10. Find the area of a stop sign with sides of 1 foot and a perimeter of 8 feet.

Solutions to Supplementary Areas of Polygons and Circles Problems

1. The sides of a square are of equal measure and the area of a square is equal to the square of a side. Before you can determine the area of the square, you must first use the perimeter to calculate the length of one side of the square. If the perimeter is the sum of the lengths of the sides, the measure of each side of the square is the perimeter divided by 4.

$$\text{Perimeter} = \frac{320}{4} = 80 \text{ ft}$$

$$\text{Area of a square} = (\text{the measure of a side})^2$$

$$A = (80)^2 (\text{ft})^2 = 6400 \text{ ft}^2$$

2. In a square, the diagonal connects opposite vertices and forms an isosceles right triangle, with legs that are two sides of the square. So, you can use the Pythagorean Theorem to find the length of each leg, which is also the length of a side of the square.

$$a^2 + b^2 = c^2$$

$$(\text{side})^2 + (\text{side})^2 = (\text{diagonal})^2$$

$$2\,(\text{side})^2 = (25)^2$$

$$(\text{side})^2 = \frac{625}{2}$$

$$\text{side} = \frac{\sqrt{625}}{\sqrt{2}} = \frac{25}{\sqrt{2}}$$

$$s = \frac{25}{\sqrt{2}} \cdot \frac{\sqrt{2}}{\sqrt{2}} = \frac{25\sqrt{2}}{2}$$

Now that you have calculated the measure of a side of the square, apply the formula for the area of a square: $A = s^2$.

$$A = \left(\frac{25\sqrt{2}}{2}\right)^2 = \frac{625 \cdot 2}{4} = \frac{625}{2} = 312\frac{1}{2}$$

3. The sides of a rhombus are congruent and the diagonals bisect each other to form four congruent right triangles with the sides of the rhombus as the hypotenuse and one-half of each diagonal as the legs of the triangle.

Apply the Pythagorean Theorem to determine the length of the unknown legs of the right triangles. Let s equal the side of the rhombus, d_1 equal the smaller diagonal, and d_2 equal the larger diagonal.

$$s^2 = (\tfrac{1}{2}d_1)^2 + (\tfrac{1}{2}d_2)^2$$

$$(12)^2 = (4)^2 + (\tfrac{1}{2}d_2)^2$$

$$144 - 16 = (\tfrac{1}{2}d_2)^2$$

$$128 = (\tfrac{1}{2}d_2)^2$$

$$8\sqrt{2} = \tfrac{1}{2}d_2$$

Therefore, the measure of the larger diagonal d_2 is equal to $2\,(8\sqrt{2})$.

$$d_2 = 16\sqrt{2}$$

4. The area of a rhombus is equal to one-half the product of the diagonals. Thus,

$$A = \tfrac{1}{2}(d_1 d_2) = \tfrac{1}{2}(8)\,(16\sqrt{2}) = 64\sqrt{2}$$

5. The area of a trapezoid is equal to one-half the product of the height and the sum of the bases. Before you can calculate the area, however, you have to determine the height of the trapezoid. As you did in the trapezoid in Example 5 of this chapter, draw two altitudes perpendicular to b_2. The constructions of the altitudes divide the figure into a rectangle and two congruent right triangles. The lower base has been divided into segments of lengths 6, 11, and 6. Use the measures of the sides and the base segment of 6 and apply the Pythagorean Theorem to calculate the height of the trapezoid.

$$(h)^2 + (6)^2 = (8)^2$$
$$(h)^2 = 64 - 36$$
$$h = \sqrt{28} = \sqrt{4 \cdot 7} = 2\sqrt{7}$$

Thus,

$$A = \tfrac{1}{2}(h)\,(b_1 + b_2) = \tfrac{1}{2}(2\sqrt{7})\,(11 + 23) = 34\sqrt{7}$$

6 through 9.

	Prob. 6	Prob. 7	Prob. 8	Prob. 9
Radius	21	9	$5\sqrt{3}$	7
Circumference	42π	18π	$10\sqrt{3}\pi$	14π
Area	441π	81π	75π	49π

Use the formulas for both the circumference and the area of a circle to determine the values in the chart. The circumference of a circle is equal to the product of twice the radius (the diameter) and the value of π: $C = 2\pi r$. The area of a circle is equal to the product of the value of π and the square of the radius: $A = \pi r^2$.

10. A stop sign is a regular octagon, and in this problem each of the sides measures 1 foot. To determine the area of the sign, you must first find the measure of the apothem, which is the perpendicular distance from the center of the polygon to a side. Each of the eight equilateral triangular segments of the sign is congruent, and the height of each (the apothem) equals $\sqrt{3}/2$. Thus,

$$A = \tfrac{1}{2}aP = \frac{1}{2}\left(\frac{\sqrt{3}}{2}\right)(8) = 2\sqrt{3}$$

Chapter 11

Volumes of Solids

Solid figures such as prisms, pyramids, cones, cylinders, and cubes require that you shift attention to the three-dimensional world from the one-dimensional world of the point and line or the two-dimensional world of the plane in which we have been working. Adding depth to figures requires calling upon additional skills in visualizing these shapes in order to correctly select our means of measuring their volumes.

In contrast to one- and two-dimensional figures, actually visualizing solids is easier because we can find numerous examples of these solids in our daily lives. Cans that contain food, hair care products, and other common items are cylindrical in shape. Cones appear in our food, on the roadway, and in the form of party hats and other objects. Children and athletes use spheres in their games of baseball, basketball, and soccer. The pyramids in Egypt are part of every child's geography lessons and occupy the interest of many television documentaries. For the most part, geometric solids most closely approach the real world of experience.

Definitions to Know

Altitude of a prism. A line segment that is also the height, which joins the two base planes of the prism and which is perpendicular to both, as in Fig. 11-1.

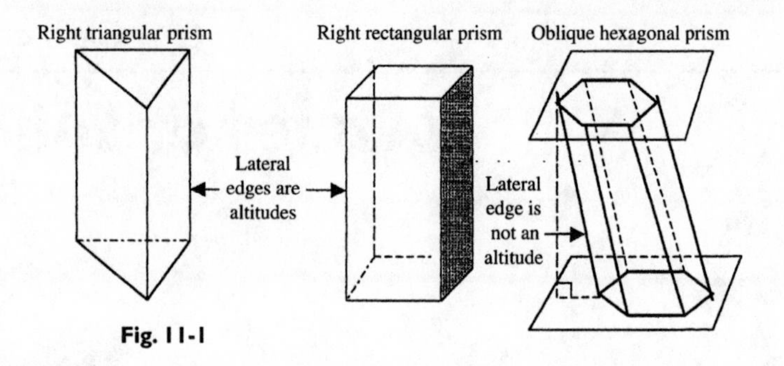

Fig. 11-1

Altitude of a pyramid. A line segment perpendicular to the base drawn from the vertex of the pyramid. Its length is the height of the pyramid, identified as h in Fig. 11-2.

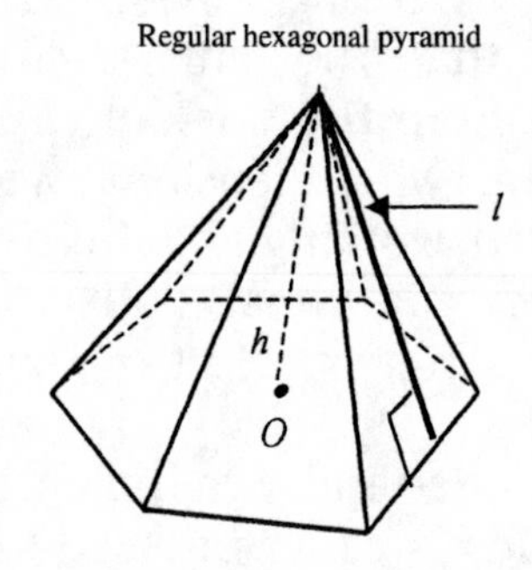

Fig. 11-2

Altitude of a right cylinder. The line segment that joins the centers of both circular bases and which is perpendicular to both, as in Fig. 11-3.

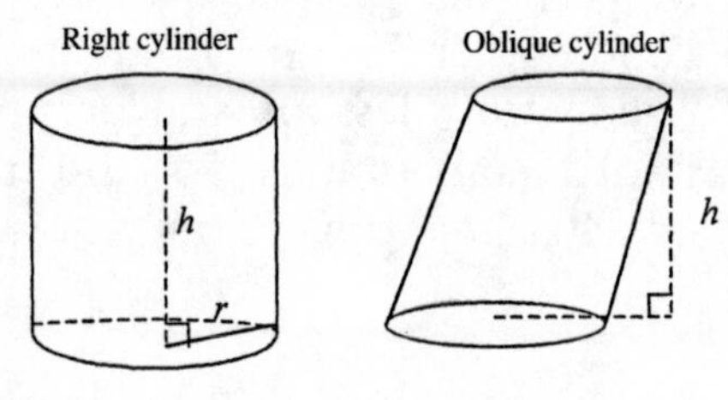

Fig. 11-3

Bases of a prism. Two congruent polygons that lie in parallel planes, as in Fig. 11-1, which give their name to individual prisms.

Cone. A shape similar to a pyramid with the exception that its base is a circle instead of a polygon, as in Fig. 11-4.

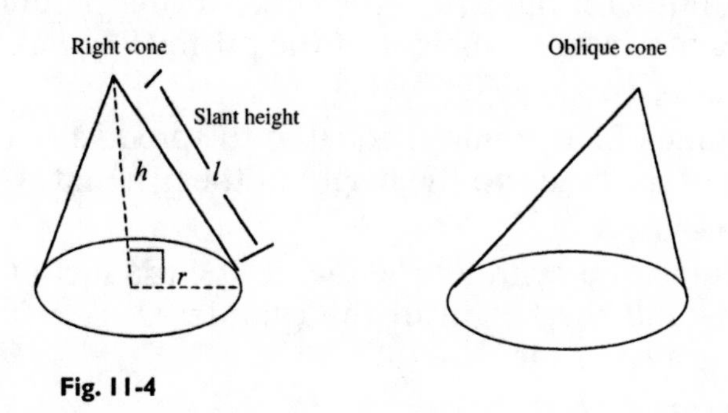

Fig. 11-4

Cylinder. A shape similar to a prism with the exception that its bases are circles instead of polygons, as in Fig. 11-3.

Lateral faces of a prism. The faces of a prism that are not its bases. Adjacent lateral faces intersect in parallel segments called lateral edges.

Prism. A solid figure that contains two bases and three or more lateral faces. The shape of the base gives the prism its name, i.e., the right triangular, right rectangular, and oblique hexagonal prisms in Fig. 11-1.

Regular pyramid. A shape that has a regular polygon for a base, congruent lateral edges, and lateral faces that are congruent isosceles triangles.

Slant height of a pyramid. The height of a lateral face, identified as *l* in Fig. 11-2.

Sphere. The set of all points that are a given distance from a given point, as shown in Fig. 11-5.

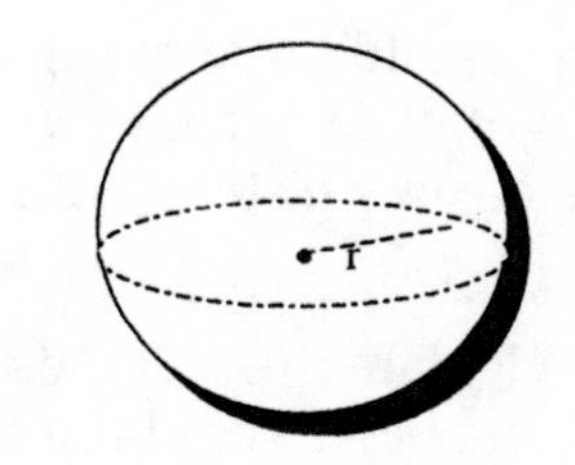

Fig. 11-5

Relevant Theorems

Theorem 1

The volume of a right prism is equal to the product of the area of a base and the height of the prism ($V = Bh$).

Theorem 2

The volume of a pyramid is equal to the product of one-third the area of the base and the height of the pyramid ($V = \frac{1}{3}Bh$).

Theorem 3

The volume of a cylinder is equal to the product of the area of a base and the height of the cylinder ($V = \pi r^2 h$), where πr^2 = the area of the base (B).

Theorem 4

The volume of a cone is equal to the product of one-third the area of the base and the height of the cone ($V = \frac{1}{3}\pi r^2 h$), where πr^2 = the area of the base (B).

Theorem 5

The volume of a sphere is equal to the product of $\frac{4}{3}\pi$ and the cube of the radius of the sphere ($V = \frac{4}{3}\pi r^3$).

EXAMPLE 1

Find the volume in terms of π of a rectangular solid such as a brick with a length of 7 inches (in), a width of 4 inches, and a height of 4 inches.

Solution

In essence, this is a rectangular prism, so apply the formula for volume that appears in Theorem 1, which states that the volume is equal to the product of the area of a base (B) and the height of the prism.

$$V = Bh = (4 \text{ in} \cdot 7 \text{ in})\ (4 \text{ in}) = 112 \text{ in}^3$$

EXAMPLE 2

The model of a pyramid has a hexagonal base with sides of 8 inches and a height of 15 inches. What is the volume of the figure? Leave the answer in terms of π.

Solution

Before you can calculate the volume of the pyramid, you must find the area of the base. Recall the formula for the area of a polygon that you used in Chap. 10. Divide the hexagon into six equilateral triangles. Find the area of one equilateral triangle and multiply it by 6.

Each triangle formed in the base of the pyramid has a base and sides of length 8; thus the height of each triangle is $4\sqrt{3}$. The area of each triangle is calculated by applying the following formula: Area of a triangle = $\frac{1}{2}bh$.

$$A = \tfrac{1}{2}(8 \text{ in})(4\sqrt{3} \text{ in}) = 16\sqrt{3} \text{ in}^2$$

The area of the base (B) equals six times the area of one triangle.

$$B = 6\,(16\sqrt{3} \text{ in}^2) = 96\sqrt{3} \text{ in}^2$$

Now substitute the value into the formula for the volume of a pyramid, found in Theorem 2: Volume of a pyramid = $\frac{1}{3}Bh$.

$$V = \frac{1}{3}(96\sqrt{3} \text{ in}^2)\,(15 \text{ in}) = 990\sqrt{3} \text{ in}^3$$

EXAMPLE 3

To fill a cylindrical tank with water, town planners must determine its volume. The tank has a base of radius 20 feet and a height of 40 feet. What is the volume of the tank? Leave the answer in terms of π.

Solution

The formula used to calculate the volume of a cylinder requires that you multiply the area of the base (B) by the height (h) of the cylinder: $V = Bh$. The base of a cylinder is a circle, so you must first determine the area of the base. As you recall from Chap. 10, the area of a circle is equal to the product of the value of π and the square of the radius: $A = \pi r^2$. The area of the base of this cylinder is as follows:

$$A = \pi\,(20 \text{ ft})^2 = 400\pi \text{ ft}^2$$

Now substitute that value in the formula used to determine the volume of a cylinder:

$$V = Bh = (400\pi \text{ ft}^2)\,(40 \text{ ft}) = 16{,}000\pi \text{ ft}^3$$

EXAMPLE 4

A cone-shaped paper cup is used to measure the amount of water that is needed to fill a pitcher. The radius of the drinking edge of the cup is 2 inches and the height of the cup is 4 inches. What is the volume of water that each cup will hold? Leave the answer in terms of π.

Solution

The volume of water that the cone-shaped cup holds can be determined by applying the formula found in Theorem 4, which states that the volume of a cone is equal to the product of one-third the area of the base (in this case, the opening) and the height, 4 inches in this problem. The area of the base is the area of a circle:

$$A = \pi r^2 = \pi \ (2 \text{ in})^2 = 4\pi \text{ in}^2$$

Now, substitute that value in the formula for the volume of a cone:

$$V = \tfrac{1}{3}\, Bh = \tfrac{1}{3}\ (4\ \pi \text{ in})^2\ (4 \text{ in}) = \tfrac{16}{3}\pi \text{ in}^3$$

EXAMPLE 5

A plane passes through the center of a sphere and forms a circle with a radius of 21 feet. What is the volume of the sphere?

Solution

The volume of a sphere can be calculated by applying Theorem 6, which states the volume of a sphere is equal to the product of $\frac{4}{3}\pi$ and the cube of the radius of the sphere ($V = \frac{4}{3}\pi r^3$).

$$V = \frac{4}{3}\pi\ (21 \text{ ft})^3\ \frac{4}{3}\pi\ (9261 \text{ ft}^3) = 12{,}348\pi \text{ ft}^3$$

Supplementary Volumes of Solids Problems

1. Find the volume of a sphere with a diameter of 12 cm.
2. A regular square pyramid has a base edge of 25 and a height of 20. What is the volume of the pyramid?
3. A cylinder has a radius of 6 and a height of 14. Find the volume of the cylinder, using 22/7 as the value of π.

4. What is the volume of a regular triangular pyramid with a base edge of 6 and a height of 8?
5. A sphere has a radius of 21 cm. Use 22/7 as the value of π to find the approximate volume of the sphere.
6. A cube with a side of $3x$ meters in length has a volume of 1728 m^3. What is the value of x?
7. A cone has a height of 30 m and a volume of 990 m^3. What is the radius of the base? Use 22/7 as the value of π to find the approximate value of r.
8. What is the ratio of the volume of a cone to the volume of a cylinder, if both have a radius of 8 and a height of 10?

Solutions to Supplementary Volumes of Solids Problems

1. The formula for the volume of a sphere is: $V = \frac{4}{3}\pi r^3$. We are given the diameter of the sphere as 12 cm, but we need the measure of the radius, so we must divide the diameter in half. Thus, the radius equals 6 cm. Substitute the value for the radius into the formula:

$$V = \frac{4}{3}\pi\,(6)^3 = 288\pi \text{ cm}^3$$

2. The volume of the pyramid is equal to the product of one-third the area of the base and the height.

$$V = \tfrac{1}{3}\,Bh = \frac{(25)^2\,(20)}{3} = \frac{12{,}500}{3} = 4166\tfrac{2}{3}$$

3. The volume of a cylinder is equal to the product of the area of the base and the height.

$$V = Bh = \pi r^2 h = \left(\frac{22}{7}\right)(6^2)(14) = 1584$$

4. The volume of a regular triangular pyramid is equal to one-third of the product of the area of the base and the height. The base is an equilateral triangle with a side of length 6, so the area of the triangle is equal to one-half the product of the base and the height.

$$A = \tfrac{1}{2}\,(6)(3\sqrt{3}) = 9\sqrt{3}$$

Now substitute the value for the area in the formula for the volume.

$$V = \frac{1}{3}(9\sqrt{3})(8) = 24\sqrt{3}$$

5. Apply Theorem 5, which states that the volume of a sphere is equal to the product of $\frac{4}{3}\pi$ and the cube of the radius of the sphere. Thus,

$$V = \frac{4}{3}\left(\frac{22}{7}\right)(21)^3 = 38{,}808 \text{ cm}^3$$

6. A cube is a right prism with equal side measures, so the volume of a cube is equal to the product of the area of the base and the height, or the side cubed. Thus,

$$V = s^3$$
$$1728 \text{ m}^3 = (3x)^3$$
$$1728 \text{ m}^3 = 27x^3$$
$$64 \text{ m}^3 = x^3$$
$$4 \text{ m} = x$$

7. The volume of a cone is equal to the product of one-third the area of the base and the height of the cone: $V = 1/3\ Bh = 1/3\ \pi r^2 h$.

$$990 \text{ m}^3 = \frac{1}{3} B(30 \text{ m})$$
$$99 \text{ m}^2 = B \quad \text{(the area of the circular base)}$$

Now solve for the radius. The area of a circle is equal to πr^2, so substitute, as below.

$$B = \pi r^2$$
$$99 \text{ m}^2 = \left(\frac{22}{7}\right) r^2$$
$$\frac{693}{22} = r^2$$
$$\sqrt{\frac{693}{22}} = r$$
$$\frac{3\sqrt{14}}{2} = r$$

8. The volume of a cylinder is equal to the product of the area of a base and the height of the cylinder. The volume of a cone is equal to the product of one-third of the area of the base and the height of the cone.

$$\frac{V_{cone}}{V_{cylinder}} = \frac{\frac{1}{3}\pi r^2 h}{\pi r^2 h} = \frac{1}{3}$$

Chapter 12

Miscellaneous Problem Drill

Leave answers in terms of π unless indicated otherwise.

1. If two angles of a triangle measure 54° and 68°, then the measure of the third angle is ______.
2. A triangle has one right angle and two equal angles. What are the measures of the equal angles?
3. If two angles of a triangle measure 107° and 43°, then the third angle measures ______.
4. In a right triangle, if one side measures 25 and a second measures 15, what is the measure of the third side?
5. If $\measuredangle A$ and $\measuredangle B$ are complementary, and $\measuredangle A$ measures 68°, what is the measure of $\measuredangle B$?
6. If $\measuredangle X$ and $\measuredangle Y$ are supplementary, and $\measuredangle Y$ measures 68°, what is the measure of $\measuredangle X$?
7. What is the complement of a 74.5° angle?
8. What is the measure of the angle formed when the hands of a clock are at exactly 3 o'clock?
9. At what time are the hands of the clock parallel to each other?
10. What angle is formed between the hands of a clock when the time is exactly 2 o'clock?
11. What is the perimeter of a rectangle of length 20 feet and width 5 feet?
12. What is the perimeter of a triangle with sides measuring 11, 15, and 17?
13. An engineer wants to construct a fence around a piece of land shaped as a regular hexagon with sides measuring 12 meters. How many meters of fencing materials are needed?

14. How many yards of fencing are needed to enclose an area that is 27 feet wide and 48 feet long?

15. The area of the floor of a square room is 169 square feet. If a carpenter wants to purchase exactly the correct number of feet of molding to place around the base of the wall, how many feet are needed?

16. A company wants to manufacture covers for round swimming pools that are 20 feet in diameter. What is the surface area of a cover if it fits exactly across the top of a pool?

17. A purchasing agent in a toy company wants to purchase boxes that will be the exact size in which to place two triangular wooden blocks, each with sides measuring 12, 16, and 20, and a thickness of $\frac{1}{2}$ inch. What are the length and width of the smallest box that can be used if the only boxes available are $\frac{1}{2}$ inch thick?

18. A square with an area of 400 square yards would require a fence of how many yards to enclose?

19. A homeowner plans to replace a kitchen floor measuring 13 feet by 24 feet with floor tiles, each measuring 1 foot by 1 foot. How many floor tiles are needed to cover the floor completely?

20. Fencing materials are sold by the yard, but a homeowner only knows that the yard that must be enclosed measures 25 feet by 29 feet. How many yards must the owner buy to make certain that the land will be completely enclosed?

21. If the measures of three angles of a triangle are 55°, 30°, and $(3x + 50)°$, then $x =$ ______?

22. If diagonals are drawn in a rectangle with sides of lengths x and y, and the vertical angles opposite the y sides are 110° each, what are the measures of each of the vertical angles opposite the x sides?

23. Find the area of a rhombus with diagonals of lengths 6 and 5.

24. Find the area of an isosceles trapezoid with legs of length 8 and bases of lengths 12 and 16.

25. What is the area of a square with a diagonal measuring 8?

26. A cylinder has a height equal to the radius of its base. If the radius of the base is 7, what is the volume of the cylinder?

27. An angle and its complement have measures of $x + 43$ and $2x - 13$. Find the measure of the angle.

28. The difference between the measures of two supplementary angles is 44°. Find the measure of each angle.

29. Find the volume of a cone with a radius of 2 and a height of 6.

30. What is the volume of a rectangular prism with a height of 10, a base of length 7, and a width of 4?

31. If the volume of a cube is 9261 cubic meters, what is the area of one face?

32. What is the volume of a cylinder with a radius of 4 millimeters and a height of 8 millimeters?

33. If $\triangle ABC \cong \triangle XYZ$, $\measuredangle A = 46°$, and $\measuredangle C = 67°$, what is the measure of $\measuredangle Z$?

34. If scientists want to build a laboratory shaped like a sphere, but must fit it into a space that is no more than 70 feet wide and high, what will be the largest radius of the sphere?

35. A carpenter needs to reach the top of a house that is 20 feet from the ground using a ladder that is 25 feet in length. How far from the base of the house must the carpenter place the bottom of the ladder so that the top of the ladder will meet exactly with the top of the house?

36. The central angle of a circle measures 48°. What is the measure of the arc formed by the intersection of the angle with the circle?

37. What is the measure of a major arc of a central angle measuring 75°?

38. What is the measure of an inscribed angle if the measure of the intercepted arc is 120°?

39. Two chords intersect in a circle, dividing chord $\overline{AB}$ into segments of lengths $3x$ and 7 and chord $\overline{CD}$ into segments of lengths 9 and 21. What is the value of x?

40. What is the area of a regular octagon with sides of length 8 and an apothem of length $4\sqrt{3}$?

41. What is the area of a right trapezoid with bases of lengths 16 and 22, and sides of lengths 8 and 10?
42. How many squares with sides of length 3 can you fit into the area of a rectangle with sides of lengths 18 and 27?
43. If a book distributor wants to save on the cost of shipping multiple copies of the same book in cardboard cartons, what is the minimum volume of a carton needed to hold 40 copies of a book having a width of 6 inches, length of 9 inches, and thickness of 2 inches?
44. What is the side measure of a hexagon inscribed in a circle having a diameter of 14?
45. How many rings measuring 6 feet in radius can a circus fit onto a floor measuring 36 feet wide and 72 feet long?
46. If the sum of two angles of a rhombus is 190°, then the sum of the remaining two angles is ______.
47. What volume of water must be used to fill a cylindrical pool with a height of 8 feet and a diameter of 40 feet?
48. A skater wants to find how many more feet she is skating each day if she skates seven times around the perimeter of a circular rink than if she simply skates 10 round trips on the diameter of the rink, which is 40 feet. Use 22/7 to approximate the value of π.
49. If the perimeters of two circles are in a ratio of 4:1, then their radii are in a ratio of ______.
50. If the volume of a cube is 1331 cm^3, what is the area of a side?

ANSWERS

1. 58°
2. 45°, 45°
3. 30°
4. 20
5. 22°
6. 112°
7. 15.5°
8. 90°
9. Noon or midnight (both hands are on the 12)
10. 60°
11. 50 feet
12. 43

13. 72 meters
14. 150 feet
15. 52 feet
16. 100π
17. 12 by 16
18. 80 yards
19. 312 tiles
20. 242 yards
21. 15
22. 70°, 70°
23. 15
24. $28\sqrt{15}$
25. 32
26. 343π
27. 63°
28. 112°, 68°
29. 8π
30. 280
31. 441 m^3
32. 128π m^3
33. 67°
34. 35 feet
35. 15 feet
36. 48°
37. 285°
38. 60°
39. 9
40. $128\sqrt{3}$
41. 152
42. 54
43. 4320 in^3
44. 7
45. 18
46. 170°
47. 3200π ft^3
48. 80 feet
49. 4:1
50. 121 cm^2

Index